Strategic Human Resource Management at Tertiary Level

Innovation and Change in Education — A Cross-cultural Perspective

Volume 4

Series Editor

Prof. Dr. Xiangyun Du
Aalborg University, Denmark

Nowadays, educational institutions are being challenged when professional competences and expertise become progressively more complex. This is mainly because problems are more technology-bounded, unstable and ill-defined with the involvement of various integrated issues. To solve these problems, it requires interdisciplinary knowledge, collaboration skills, innovative thinking among other competences. In order to facilitate students with the competences expected in professions, educational institutions worldwide are implementing innovations and changes in many aspects.

This book series includes a list of research projects that document innovation and change in education. The topics range from organizational change, curriculum design and innovation, pedagogy development, to the role of teaching staff in the educational change process, and quality issues, among others. A cross-cultural perspective is studied in this book series that includes two layers. First, research contexts in these books include different countries with various educational traditions, systems and societal backgrounds. Second, the impact of professional and institutional cultures such as engineering, medicine and health, and teachers' education are also taken into consideration in these research projects.

For a list of other books in this series, visit www.riverpublishers.com

Strategic Human Resource Management at Tertiary Level

Dr. M. D. Tiwari

Director IIIT
Allahabad & Amethi &
Former President
Association of Indian Universities
New Delhi
India

Dr. Iti Tiwari

Associate Professor
Uttar Pradesh Rajarshi Tandon Open University
Allahabad
India

Dr. Seema Shah

Deputy Registrar
Indian Institute of Information Technology
Allahabad
Uttar Pradesh
India

River Publishers

Routledge
Taylor & Francis Group

LONDON AND NEW YORK

Published 2013 by River Publishers
River Publishers
Alsbjergvej 10, 9260 Gistrup, Denmark
www.riverpublishers.com

Distributed exclusively by Routledge
4 Park Square, Milton Park, Abingdon, Oxon OX14 4RN
605 Third Avenue, New York, NY 10158

First published in paperback 2024

Strategic Human Resource Management at Tertiary Level / by M. D. Tiwari, Iti Tiwari, Seema Shah.

Routledge is an imprint of the Taylor & Francis Group, an informa business

Publisher's Note
The publisher has gone to great lengths to ensure the quality of this reprint but points out that some imperfections in the original copies may be apparent.

While every effort is made to provide dependable information, the publisher, authors, and editors cannot be held responsible for any errors or omissions.

ISBN: 978-87-92982-58-2 (hbk)
ISBN: 978-87-7004-504-9 (pbk)
ISBN: 978-1-003-35722-3 (ebk)

DOI: 10.1201/9781003357223

Editors' Biography

Dr. M D Tiwari is the founder Director of the Indian Institute of Information Technology Allahabad and Rajiv Gandhi Institute of Information Technology Amethi is recognized for his excellent academic and administrative achievements in Education, Research, Science, Technology, Management and Training.

Dr. Tiwari is an acclaimed Institution Builder. As Head developed Physics department at Gharwal University; Bureau Chief in University Grants Commission, COSIST, S&T, he initiated Emerging Areas Schemes, Operated and Supervised these nationwide. As Advisor in AICTE he is known for his path breaking work in Thrust Areas, R&D, International Collaboration and Administration. As Vice-Chancellor or Rohilkhand University Dr. Tiwari established institutes (E&T, Management, Educational Sciences); launched several professional courses; academic sessions normalized; and ensured that the University became financially strong. During his tenure the University was adjudged first in Performance Appraisal for consecutive three years amongst all Universities of U.P. and in recognition State Government awarded for Rs. 4 crores.

As first Chairman of the UP State Council for Higher Education, Dr. Tiwari operationalized several innovative schemes. Dr. Tiwari headed Committee on Professional Education Courses for streamlining and rationalizing fee Structure. This is implemented by UP State. Under Dr. Tiwari's Chairmanship detailed Project Report including the Act, Statues and Ordinances were prepared for UP State Open University. Member the Act, Statues and Ordinances were prepared for UP State Open University. Member for Project Report of Indian Institute of Information Technology and established it at Allahabad ad it's O.S.D (First C.E.O).

Dr. Tiwari is recipient of several fellowships and awards. National Nodal Expert for Indo-Swiss and Indo-Canada S&T (ICT) Programme. Chairman of several Committees of DST, AICTE and other Agencies, Published over 165 Papers in reputed journals and Books, Supervised over 10 Ph.Ds.

Dr. Iti Tiwari is Associate Professor in Uttar Pradesh Rajarshi Tandon Open University, Allahabad, India. Prior to that she was Reader & Head of Department Sociology at Jagadguru Rambhadracharya Handicapped University at Chitrakoot, Faculty in Rakshapal Bahadur Institute of Management at Bareilly and coordinator (Distance Education) G.G.D. University at Bilaspur. Her specialization is in child crime and Handicapped. She has tremendous contributions in the area of physically challenged persons and child crime. She has published more than 150 articles and manuscripts, 3 books and substantial research publications.

Dr. Seema Shah has been awarded D.Phil degree from the famous University of Allahabad, U.P., India in the area of Human Resource Management. She is crowned with gold medal of being topper of P.G. in Human Resource Management of M.P. Bhoj Open University, Bhopal, India. She has already acquired Master Degree from Kumaun university, Nainital. Her Knowledge of Application of ICT in Human Resource Management and other related matters have been par-excellence elsewhere by developing new softwares. She has already published several research papers in International journal of world repute, which have been copyrighted by Microsoft and Google Scholar as well. They have been cited by several authors globally. She has also published an edited book and organized several National/ International Conferences/ Seminars. Her interest is in the Application of ICT in HRM of Educational Institutions. She is working in Indian Institute of Information Technology, Allahabad, U.P., India as a Deputy Registrar.

Contents

7   Developing Strategic International Human Resource Management: Prescriptions for MNC Success   **103**

Mary Ann Von Glinow, John Milliman

8   Development of a Knowledge Management Framework within the Systems Context   **137**

Roberto Biloslavo, Max Zornada

9   HR Practices, Social Climate, and Knowledge Flows: Towards Social Resources Management   **165**

Angelos Alexopoulos, Kathy Monks

Preface

The modern Human Resource Management practices using Information Technology tools are one of the valuable aspects at international level. Now Electronic Era (e-era) has further generated challenges and interest in this area. The key responsibilities of HRM (IT) practices are planning, setting goals, assessing proper selections and interpersonal relation, team working, watching interest of teachers and employees and transparent system of management.

The first chapter uses e-learning tools as a support for traditional classroom teaching started at the University of Rijeka University at the beginning of 2008. The results from general importance of specific e-learning elements as a part of the survey indicates that the students value is the most completeness, organization and design of educational materials, as well as teacher's online engagement especially in good management of e-course in regular communication and timely providing feedback.

The second chapter presents the revolution in information technology helping in ensuring effectiveness of HR functions. Human Resource Information System (HRIS) is an opportunity for organizations to make the HR department administratively and strategically participative in operating the organization. The holistic view of the role that HRIS can play in improving the efficiency and integration of HR department into a more strategic role was missing. Neither cost saving or strong communication nor effective recruitment decisions were linked directly with HRIS.

The third chapter discuss many centuries India has absorbed managed ideas and practices from around the world. The historical and traditional roots remain deeply embedded in the world in a subjective manner, emphasis on objective global concepts and practices are becoming more common. The nature of hierarchy, status, authority, responsibility and similar other concepts vary widely across the nations synerging system maintenance, indeed organizational performance and personal success are critical in the new era.

The fourth chapter shows a brief overview about possibilities of IT usage in HR field for measuring and tracking human capital and using the HR information system. Generally, Globalization brings the requirement to think how IT can contribute to fulfillment of business strategy in the frame of Human

Resource management in order to steer the business towards excellence and manage competitiveness in the market.

Fifth chapter discusses the findings of the needs analysis in SE Europe and to reflect upon them through cultural diversity perspective and marketization point of view. Higher education institutions focus their efforts on the international recognition and comparability and design programmes that can best meet the needs of international students.

The sixth chapter provides an individual level approach of the human resource management requiring focus upon the employees' individual development within the organization. As a results similar development needs assessments which may be extremely useful in designing and investing in successful human resources improvement programs, in order to the institutional development programs to distinctly address the employees subgroup.

The seventh chapter shows that U.S. multinational corporations (MNCs) have experienced difficulties in their overseas operations, due in part to ineffective international human resource management (HRM) practices. This paper uses a product life cycle (PCL) approach to develop a two-step contingency model of the strategic and operational levels of MNCs. In this chapter toward enhancing MNC success through IHRM practices can also serve as a research agenda for scholars and practitioners alike.

In the eight chapter proposed framework is particularly focused on dividing the identified organizational building blocks into their constituent elements along with both time and content dimensions so as to define characteristics that these elements and the relationships between them need to have to form a social ecology in which people effectively create share and use knowledge.

The ninth chapter suggests a strong linkage between HR Systems and knowledge management outcomes. The study contributes to a better understanding of the breadth of HR systems in a knowledge-intensive organizational context. Finally in terms of depth, the study suggests that the effective management of social relations may require a process-based HR approach that goes beyond explicit motivation mechanisms, such as pay incentives, for sharing knowledge, and directs attention to core structural aspects of knowledge work as well as to softer incentives for supporting prosocial behaviours and value-creating social relations.

The tenth chapter aims at analyzing the relationship between innovation and human resource management (HRM) attempting to establish whether innovation determines the firm's human resource management. The use of HRM practices aimed at building a stable group of employees in the company, which can adopt risks and experiment and which can participate in the adoption of the decisions that affect their jobs. This is more likely to create the conditions for the emergence of the new ideas that feed innovation.

The eleven chapter describe Human resources management is one of the departments that mostly use management information systems. HR information systems support activities such as identifying potential employees, maintaining complete records on existing employees and creating programs to develop employees' talents and skills. Thus future studies should also consider the relationships between the access limitations to information content and functions of HRIS and user satisfaction. Overall the contribution provides valuable insights into the study of HRIS success.

The twelfth chapter discusses an individual level approach of the human resource management which requires the focus upon the employees' individual development within the organization. As regards the limits of the present work, in order to immediately meet the organizational daily needs, one of the main difficulties is generated by the lack of organizational long term strategies and objectives.

The Editors' are extremely grateful to authors' who has been a great support for this work-Martz Zuvic-Butorac, Nena Roncevic, Damir Nemcanin, Zoran Nebic, Usman Sadiq, Ahmad Fareed Khan, Khurram Ikhlaq, Bahaudin G. Mujtaba, Prof. Samir R. Chatterjee, Ing. Iveta Gabcanova, Dr. Anita Trnavcevic, Dr. Nada Trunk Sirca, Mag. Vinko Logaj, Chukwunonso Franklyn, Mary Ann Von Glinow, John Milliman, Roberto Biloslavo, Max Zornada, Angelos Alexopoulos, Kathy Monks, Professor Marius Dan Dalota, Yasemin Bal, Serdar Bozkurt, EsIn Ertemsir, Osoian Codruta.

The Editors are also thankful to concerned Journals for their support. Sincere acknowledgement are also extended by Editors to Journals of Information Science & Information Technology-USA, Journal of Business Studies Quaterly, Journal of Research and Practice in Human Resource Management-Australia, 6th International Conference 'Managing the process of Globalisation in New and upcoming EU Members'of the Faculty of Management Koper Congress Centre, Bernardin, Slovania, Centre for effective Organization (CEO), Marshall School of Business, University of Southern California, Los Angeles, Link Working paper Series, International Conference, 2012 'Management Knowledge and learning. Authors of the papers are responsible for plagiarism and other related aspects in the book. The Editors have no role in this regard.

Editors

1

Blended E-Learning in Higher Education: Research on Students' Perspective

Marta zuvic- Butorac[1], Nena Roncevic[2], Damir Nemcanin[1]
and Zoran Nebic[1]

[1]*Faculty of Engineering, University of Rijeka, Rijeka, Croatia*
[2]*Faculty of Humanities and Social Sciences, University of Rijeka,
Rijeka, Croatia*

1.1 Abstract

The process of implementing blended learning, by using e-learning tools as a support for traditional classroom teaching, started at the University of Rijeka at the beginning of 2008, following general strategic principles adapted to local environment. The process has been constantly supported and assessed for quality, but up to now only from the institutional, teachers' and support services' perspective. Assuming that continuous and careful monitoring of learner's satisfaction is important for the success, feasibility and viability of e-learning, we conducted the research on students' perspective. As the student's perception regarding e-learning is one of the most important steps in developing and implementing a successful e-learning environment, we conducted the study of student's perception and e-learning acceptance, with three specific items addressed in the study: 1) student's perception of quality of already delivered e-courses, 2) level of importance for the specific elements of e-learning encountered, and 3) student's general attitude towards e-learning and their needs with respect to quality of course materials, communication and support of the learning process.

Participants in the study assessed the current state of e-learning elements implementation quite good; they agreed the educational materials were in most cases complete, organized and well designed, and they considered the ability of teachers to manage the e-courses well, communication regular as well as

1

the provision of the feedback. The lower level of agreement was obtained on the use of multimedia, offering of the self-assessment tests, accessibility of digital literature and collaborative activities. This suggests teachers should be encouraged and trained to put more effort in designing and offering suitable multimedia elements to enrich their materials, self-assessment test to make students feel more comfortable in terms of examination expectations, and to design online activities for the students to enhance collaborative aspects in teaching. The results obtained from "the general importance of Specific e-learning elements" part of the survey indicated that students value the most the completeness, organization and design of educational materials, as well as teachers' online engagement, especially in good management of e-course, in regular.

Communication and timely providing feedback. They do not perceive as much as important the online activities, communication to other students and discussions. When the comparison of "current state" and "general importance" for the specific e-learning elements is made, it seems that there is not much of discrepancy.

Through assessment of the general value of e-learning and its characteristic, participants best agree with the notion that most important is to have the access to teaching materials 24/7. Second best is that online materials are better suited to students' needs and that in general, having e-course as addition to classroom teaching is helping them organizing their learning better and scoring better results.

As the general attitude towards online learning is considered, it is interesting that preferences for exclusively online and/or blended learning are dominant in a group of students having better average studying grades (A or B), as well as in a group having shorter studying experience (1–2 yrs. of studying). The same groups find the new communication channels (online discussion forums, mailing with teachers, assistants and colleagues within the online learning environment) important and useful. Moreover, they think that online educational materials and activities are better suited to students' needs and that they help them achieve the learning outcomes better. This result is an important signal to all instances supporting e-learning implementation.

The study outcomes generally suggest the need to enhance teachers' competencies for online teaching, particularly in acquiring successful tutoring methods and learners' support methods, and together with continuous and careful monitoring of learner's satisfaction we hope to ensure the success, feasibility and viability of online learning, as supporting educational tool in our university study programs.

Keywords: Blended learning higher education, Learner's perspective, Quality assurance, E-learning acceptance.

1.2 Introduction

As education is becoming a ubiquitous service delivered anywhere and any-time over the global network, the higher education institutions, although campus oriented and without distance learners, try to implement elements of e-learning in traditional course delivery, in order to prepare their students, as well as the institution, for the future participation in education (Bonk, 2009; McCradie, 2003). In this process, there is also a hope that such changes will also induce some changes in traditional organization, planning and management of educational process.

University of Rijeka is one of the seven universities in Croatia, middle-sized with respect to number of students (∼17 500) and academics (∼1 100). As dynamic and change-oriented institution, in its policy documents in 2007 the University defined the strategic goals, particularly related to teaching and learning (T&L) process and improvement of its overall quality. Thus the specific strategic objectives were set up: to increase the efficiency of studying, to modernize curricula and syllabi in the context of the Bologna declaration, to ensure the compatibility with international educational systems, to improve quality of teaching and learning through implementation of learning-outcomes oriented curricula, More over the goal was to increase the inter-university and international cooperation and to enhance the student and teacher mobility and to improve student services. Additionally motivated by the poor use of ICT in teaching and learning process, together with changes in curricula mandated by the Bologna process, the University management decided to enable the activities for e-learning implementation (Zuvic-Butorac & Nebic, 2009). As University of Rijeka is campus based, the e-learning implementation was seen in the form of transforming pre-existing traditional classroom content delivery to combination of classroom and online delivery (blended learning), through setting up of e-courses which will support the classroom activities.

The process of implementing e-learning tools as a support for traditional classroom teaching started at the University of Rijeka at the beginning of 2008, following the strategic principles (Bates, 1999; Duderstadt, 2003; Ellis, 2007; Hanna, 2003) adapted to local environment (Zuvic-Butorac, 2009). Since the time of the beginning of e-learning use and implementation of blended learning, the process has been constantly supported (through development of support services and education of teaching staff) and assessed for quality, but only from the institutional, teacher's and support services' perspective. Assuming that quality of the teaching and learning process is not something that is delivered to a student by e-learning provider, but rather constitutes a process of co-production between the learner and learning environment, we considered equally important to asses both the learner's perspective as well as learning environment aspects. In broader sense, the learning environment nowadays and particularly with the e-learning employed, is very complex and consists of many elements which contribute to its quality. It starts from the characteristics of the e-learning platform, technological and educational user's support, course design and T&L methods and tutoring employed, all the way up to institutional support and management policies towards all participants in the educational process. Assessing the quality is therefore as much as important for the students, as for the university management, support services and the academics, as teachers, authors and tutors. Understanding that student's perception regarding e-learning is one of the most important steps in developing and implementing a successful e-learning environment (Keller & Cernerud, 2002; Wagner & Flannery, 2004), we conducted the study of student's perception and e-learning acceptance, presented here.

Three specific questions were addressed in the study: 1) student's perception of quality of already delivered e-courses, 2) level of importance for the specific elements of e-learning encountered, and 3) student's general attitude towards e-learning and their needs with respect to quality of course materials, communication and support of the learning process. Additionally, students were asked to assess the technical aspects of LMS use. The paper describes the results of the study and suggests possible implications for quality improvement.

1.3 Research Methodology

1.3.1 Sample

All the students at the University of Rijeka that were using MudRi e-learning system (Moodle open source learning management system) from February 2009 to February 2010 have been selected as a sample. The questionnaire is created as an online survey using the Lime Survey open source software (Version 1.87). A request to take part in the survey and a direct link to it, with guaranteed privacy, has been sent via email to all the students in the MudRi user database. From the 1944 requests that were sent out, 361 questionnaires were received (19% response), including 48 incomplete questionnaires which were eliminated from the further process (so the actual response rate was 16%). As the survey was completed, the answers were automatically stored in the digital database, which was later used in the statistical data analysis. Data collection took place during January and February 2010.

The total number of the analyzed questionnaires is 313. The age of the respondents is 22 ± 4 (median: 20 years of age, 10–90 percentile range: 19–24 years of age). According to gender, there are more male respondents (62 % male vs. 38 % female respondents). Regarding the type of studies, 86 % of the respondents are a part of university studies while 14 % take part in professional studies. According to field of study, respondents are divided into three closely related categories: the group of studies that includes engineering, mathematics, natural sciences and medicine (ENGMATSCI, 40 % of respondents), the group involving social sciences and humanities (SSH, 18 % of respondents) and the group that includes information and communication technologies (ICT, 42 % of the respondents). Regarding experience, the majority, 39 % of students, are at their first year, 36 % have been studying for 2–3 years, and 25 % of respondents have been studying for more than 3 years. According to the number of previously used e-courses, 28 % have used up to 2 e-courses, 44 % have used 3–5 e-courses, and 28 % have used more than 5 e-courses. Regarding previous experience of e-learning, 57 % of the respondents have one year experience, while 43 % are more experienced users. Concerning success in studies, 16 % of the respondents are excellent students (grade A), 37 % are very good students (grade B), 40 % are good (grade C) and 7 % of the students are adequate (grade D).

1.3.2 Methods

Data are collected using an original questionnaire which is created for the purpose of this research. It is partly based on the experiences gained from the published researches (Bernard, Brauer, Abrami, & Surkes, 2004; Davis, 1989; DeLone & McLean, 2003; Poelmans, 2009) and partly on the experiences gained in the similar research conducted on the smaller sample including students from the Faculty of Philosophy at the University of Rijeka (Roncevic, Ledic, & Vrcelj, 2009). The questionnaire is divided into four parts. The first part includes general sociodemographic variables of the respondent (gender, age, size of place of residence, type of studies, year of study), the experience of e-learning (the number of attended e-courses, years of e-learning experience) and the preferences regarding the mode of courses (blended courses compared to direct, traditional and fully online courses). The second part of the questionnaire refers to the assessment of the existence of the certain e-learning elements in the used online courses, and the third part questions students' attitudes towards e-learning and the experience of using MudRi system. In the final, fourth part of the questionnaire, respondents assesses the personal importance of the existence of certain e-learning elements, i.e., what they do and do not find important and useful in an e-course.

The questionnaire defines 59 variables which are processed by appropriate statistical methods. The questions regarding e-course experience, attitudes towards e-learning and assessment of the importance of the e-learning elements were taken as dependent variables, and the independent variables are age and gender, size of place of residence, type and field of studies, experience of e-learning, grade point average and preferences according to course mode (classroom/blended/online).

1.3.3 Data Analysis

Data are collected electronically and automatically stored into database. The software package SPSS 18.0 was used for statistical analysis. The categorical data were described by frequencies and percentages, while numerical data by means and standard deviations. Yates corrected Pearson Chi-square test was applied to determine statistically significant differences in contingency tables of frequencies. Numerical sets of data were compared using t-test or analysis of variance, where suitable. Comparison of specific groups in analysis of variance was performed using appropriate post-hoc tests (Scheffe test for homogeneous and Tamhane T2 for non-homogeneous variance). The level of statistical significance, p, was set at 0.05 in all analyses. To determine the factor structure (latent dimensions) for a set of variables defined in specific parts of the questionnaire, factor analysis was used.

1.4 Results

1.4.1 Sociodemographic Status of the Respondents and Student Profile

Even though the first part of the questionnaire refers to the variables which are assumed to be independent, their potential relationship has been analyzed, so the conclusions could be interpreted correctly. The field of study is, according to expectations, significantly related to the gender (Pearson Chi-square: 79.36, df=2, p<0.001), where female respondents study social sciences and humanities in significantly higher number and male respondents study engineering, natural sciences and ICT. The number of e-courses that respondents use is closely connected to the field of study (Pearson Chi-square: 61.4, df=4, p<0.001), where it can be seen that engineering, social sciences and ICT students have significantly higher number of e-courses than the students of social sciences and humanities. In addition, as anticipated, there is a significant correlation between the experience of e-learning and the field of study (Pearson Chi-square: 12.5, df=4, p=0.002), where ICT students have more extensive e-learning experience than others. The attitude towards blended courses in comparison to direct, classroom and fully online courses has been analyzed. According to data, 77 % of the respondents state that they prefer blended learning, 17 % opt for exclusively classroom courses, while 7 % find the fully online courses the optimal learning and teaching mode. Preference for certain mode of courses is significantly related to the success in studies (Pearson Chi-square: 12.63, df=6, p=0.049), where remarkably higher number of students with high grade point average (very good and excellent, B and A) support blended learning, while the supporters of traditional courses are graded somewhat lower (good, C). Interestingly, the preference for certain course mode (blended/classroom/online) is not related to any of the sociodemographic or student profile variables.

1.4.2 Assessment of the Existence of E-Learning Elements in the E-Courses Using the MudRi System

The second part of the questionnaire refers to the e-course users assessment of the actual existence of certain e-learning elements in courses presented on university LMS, MudRi. The assessment was expressed by rating each statement on a five-point scale ranging from "entirely incorrect" (numerical score 1) to "entirely correct" (numerical score 5). The results are shown in the Table 1. The statement rated as the most correct by 60 % of the respondents is "E-course enabled Forum discussions". Interestingly, at the same time only 20 % of the students involved in the online courses claim to have communicated with other colleagues from the group, which means that Forums, although set by the teacher, are used rarely. Other statements were regarded mostly as partly

Table 1 Actual Existence of Certain E-Learning Elements

Statement	1	2	3	4	5	Mean	SD
I1 The e-course provided all the materials needed for achieving the expected learning results.	1	9	20	38	32	3.9	1.0
I2 Learning materials in the e-course were written in a clear and understanding manner, they were delicately colored, and had a simple standardized form.	1	4	21	38	37	4.1	0.9
I3 The learning materials and activities in the e-course were well organized.	0	7	20	41	32	4.0	0.9
I4 Multimedia (appropriate audio and video content, animations, computer simulations, etc.) was used in the e-course.	19	26	23	16	16	2.9	1.4
I5 Some educational activities in the e-course were conducted online (doing homework, submitting the seminar papers, participating in discussions, etc.).	5	8	15	31	40	3.9	1.2
I6 It was easy to communicate with teacher/assistant through the e-course.	4	7	24	30	35	3.9	1.1
I7 Through the e-course I communicated with other colleagues from the group.	18	17	25	21	20	3.1	1.4
I8 E-course enabled Forum discussions.	2	3	11	24	60	4.4	0.9
I9 E-course provided ways to test knowledge through self-assessment.	23	20	19	24	15	2.9	1.4
I10 E-course provided mandatory and optional study material in digital form.	14	23	27	22	14	3.0	1.3
I11 E-course teacher edited content and managed e-course activities regularly.	1	5	20	35	40	4.1	0.9
I12 E-course teacher used the system to communicate with students regularly.	3	7	19	35	36	3.9	1.0
I13 I regularly received feedback about my work from e-course teacher.	2	7	25	32	34	3.9	1.0

correct or entirely correct. Along these lines, 70 % of the respondents find correct or mostly correct the statements about the learning materials being written in a clear and understanding manner, that they were delicately colored, had a simple standardized form, were coherent and well organized, that some educational activities were performed online and that the teacher edited the content and managed e-course activities regularly and often used the system for communication with students. Relating to the communication coming from students, 65 % find it easy to communicate with teacher/assistant through e-course. Implementation of features specific for online learning were rated as least correct or were rarely regarded as correct. For example, the implementation of the multimedia, self-assessment tests and mandatory and optional study material (in digital form) are assessed as good in 14 to 16 % of the cases. These results are not surprising, regarding the fact that implementation of multimedia requires teachers to acquire specific knowledge and skills, and the creation of the self-assessment tests requires an extensive database of questions. Study material in digital form presents a particular problem since they are protected by copyright.

All the values of variables I1 to I13 have been tested for the differences between groups defined by independent variables. We will single out only the interesting results. In contrast to our expectations, the results show that the students of the SSH group of studies rate implementation of the multimedia, online activities and the teacher's regularity in editing content and online activities higher than other students. SSH students assess implementation of multimedia (3.3 ± 1.4 vs, 3.0 ± 1.3 ENGMATSCI and 2.5 ± 1.2 ICT students, $F=10.65$, $p<0.001$) and e-course online activities (4.4 ± 0.9 vs. 3.8 ± 1.3 ENGMATSCI and 3.9 ± 0.9 ICT students, $F=6.58$, $p=0.002$) with a significantly higher score than other students. The teacher's regularity in editing content and activities in the e-course is rated significantly lower by ICT students (3.9 ± 0.9 vs. 4.1 ± 0.9 ENGMATSCI and 4.3 ± 0.9 SSH students, $F=3.49$ $p=0.032$). Equally interesting is the result regarding the differences in rating experiences in comparison with the attitude towards the course, i.e., the preference for the course mode. Different from online course supporters and traditional classroom supporters, the students preferring blended learning describe e-learning materials as coherent and sufficient for achieving the expected learning results. On the other hand, the respondents with the attitude "classroom courses are the optimal ones" give lower scores for the organization of the learning materials and activities, the experiences with all aspects of communication, and to the teacher's involvement in the e-course (Table 2.).

Table 2 Assessment of the Existence of E-Course Elements in Relation to the Preference for the Certain Type of Learning

		I Prefer							
		Blended		Online		Classroom		ANOVA Test	
	Statement	Mean	SD	Mean	SD	Mean	SD	F	p
I1	The e-course provided all the materials needed for achieving the expected learning results.	4.0	0.9	3.7	1.0	3.6	1.1	5.03	0.007
I3	The learning materials and activities in the e-course were well organized.	4.1	0.9	4.0	1.0	3.5	0.9	8.76	<0.001
I6	It was easy to communicate with teacher/ assistant through the e-course.	4.0	1.1	4.2	1.1	3.2	1.1	11.66	<0.001
I7	Through the e-course I communicated with other colleagues from the group.	3.2	1.3	3.4	1.6	2.4	1.3	6.99	0.001
I8	E-course enabled Forum discussions.	4.4	0.9	4.7	0.7	4.0	1.0	4.84	0.009
I11	E-course teacher edited content and managed e-course activities regularly.	4.1	0.9	4.1	0.9	3.8	0.9	3.30	0.038
I12	E-course teacher used the system to communicate with students regularly.	4.0	1.0	4.0	1.0	3.6	1.0	3.30	0.038
I13	I regularly received feedback about my work from e-course teacher.	4.0	1.0	4.1	0.9	3.5	1.0	4.31	0.014

1.4.3 The Importance of the Existence of E-Learning Elements in E-Courses

This part of the questionnaire determines the importance of e-learning elements in e-courses for the students. As opposed to the assessment of actually present elements, students value the general importance of the same elements more (Table 3. vs. Table 1.). More than 90 % of the respondents find important or extremely important all the statements relating to the organization, coherence and clarity of the learning materials (S1, S2, S3) and statements relating to the teacher's activities and editing (S11, S12, S13). They give a lower score for the importance of the existence of online activities, knowledge self-assessment tests and communication with the teacher using the system (S5, S6 and S9). The least important for the students is the existence of the multimedia elements (S4), forum discussions (S8) and the ability to communicate to their colleagues through the e-course (S7).

All the statements in this part of the questionnaire were analyzed considering groups defined by independent variables. Thus, female respondents give significantly higher scores for the statements describing the coherence and the organization of learning materials as well as receiving feedback from the teacher (to all statements 4.7 ± 0.6 vs. 4.5 ± 0.7 with male respondents, t-test, $p<0.010$). According to field of study, ENGMATSCI students find good organization of the learning materials less important than other students (4.5 ± 0.8 vs. 4.7 ± 0.6 with ICT and 4.8 ± 0.5 with SSH students, ANOVA, $F=3.71$, $p=0.026$), as well as the existence of multimedia (3.5 ± 1.2 vs. 3.7 ± 1.1 with ICT and 4.1 ± 1.0 with SSH students, ANOVA, $F=4.65$, $p=0.010$). Students with shorter study experience and e-course use experience give significantly higher score for the ability to communicate with other colleagues from the group (3.7 ± 1.1 with first year students vs. 3.2 ± 1.3 with second year students and older, ANOVA, $F=6.19$, $p=0.002$) and they value the ability of forum discussions more (3.9 ± 1.0 with first year students vs. 3.4 ± 1.3 with second year students and older, ANOVA, $F=5.84$, $p=0.003$). Observing the differences in rating in relation to the attitude towards the course (preferences for blended/online/classroom), students opting for traditional classroom courses give significantly lower scores for a large number of statements (Table 4.): the coherence of the learning material, the existence of the online activities, communication with other colleagues and the teacher, Forum discussions and teacher's involvement (regularity of content and activity editing, communication with students, providing feedback).

Table 3   Importance of the Existence of Certain E-Course Elements in General

Statement	Answer / %					Mean	SD
	1	2	3	4	5		
S1 It is important to me that e-course provides all the materials needed for achieving the expected learning results.	0	1	8	23	68	4.6	0.7
S2 It is important to me that learning materials in the e-course are written in a clear and understanding manner, that they are delicately colored, and have a simple standardized form.	0	1	9	26	64	4.5	0.7
S3 It is important to me that learning materials and activities in the e-course are well organized.	0	1	5	24	69	4.6	0.7
S4 It is important to me that multimedia (appropriate audio and video content, animations, computer simulations, etc.) is used in the e-course.	5	10	28	27	30	3.7	1.2
S5 It is important to me that some educational activities in the e-course are conducted online (doing homework, submitting the seminar papers, participating in discussions, etc.).	3	5	16	31	45	4.1	1.1
S6 It is important to me that it is easy to communicate with teacher/assistant through the e-course.	5	7	23	30	34	3.8	1.1
S7 It is important to me that through the e-course I communicate with other colleagues from the group.	9	14	25	25	26	3.4	1.4
S8 It is important to me that e-course enables Forum discussions.	6	11	26	28	29	3.6	1.2
S9 It is important to me that e-course provides ways to test knowledge through self-assessment.	5	5	23	28	40	3.9	1.1
S10 It is important to me that e-course provides mandatory and optional study material in digital form.	1	3	11	24	61	4.4	0.9
S11 It is important to me that e-course teacher edits content and manages e-course activities regularly.	0	0	4	19	77	4.7	0.6
S12 It is important to me that e-course teacher uses the system to communicate with students regularly.	1	2	11	29	58	4.4	0.8
S13 It is important to me that I regularly receive feedback about my work from e-course teacher.	0	0	9	22	69	4.6	0.7

Table 4 Importance of Certain E-Course Elements in Relation to the Course Mode Preference

| Statement: "It is important to me that:" | I Prefer | | | | | | ANOVA Test | |
| | Blended | | Online | | Classroom | | | |
	Mean	SD	Mean	SD	Mean	SD	F	p
S1 E-course provides all the materials needed for achieving the expected learning results.	4.6	0.6	4.8	0.5	4.3	0.9	8.1	<0.001
S5 Some educational activities in the e-course are conducted online.	4.3	0.9	4.4	1.2	3.1	1.1	37.7	<0.001
S6 It is easy to communicate with teacher/assistant through the e-course.	3.9	1.1	4.3	0.9	3.0	1.1	16.7	<0.001
S7 Through the e-course I communicate with other colleagues from the group.	3.5	1.3	4.0	1.2	2.9	1.2	7.7	0.001
S8 E-course enables Forum discussions.	3.7	1.2	4.2	0.7	3.0	1.1	10.6	<0.001
S9 E-course provides ways to test knowledge through self-assessment.	4.0	1.1	4.2	1.1	3.4	1.1	7.3	0.001
S11 E-course teacher edits content and manages e-course activities regularly.	4.8	0.9	4.8	0.6	4.5	0.8	6.2	0.002
S12 E-course teacher uses the system to communicate with students regularly.	4.5	0.5	4.7	0.5	4.1	0.8	5.7	0.004
S13 I regularly receive feedback about my work from e-course teacher.	4.6	0.8	4.8	0.7	4.4	1.0	4.2	0.015

1.4.4 Students' Attitudes towards E-Learning and E-Courses in General

The final part of the questionnaire refers to the students' attitudes towards e-learning and e-courses in general (Table 5.). Bearing in mind that all of the courses using MudRi system at the University of Rijeka are realized as blended learning mode, these attitudes as well should be interpreted through the aforementioned prism. The statement students agree with the most is the attitude towards the high importance and utility of having unlimited access to all materials (88 % correct or entirely correct). Somewhat lower in percentage, but still highly rated, are the attitudes that the learning materials in e-courses are more suitable for the students (66 % correct and entirely correct), that e-course contributes to better learning organization (56 % correct and entirely correct), and the attitude that the blended course mode produces better results (53 % correct and entirely correct). It is also important to notice that 56 % of the respondents find face-to-face contact with the teacher important to encourage their learning process. However, a large number of students (one third to one fourth of the students) cannot rate the attitudes (except S17); for example, more than 30 % cannot assess whether they achieve better results in blended mode course, their level of activity in e-courses and their communication with teacher and other students. These results indicate insufficient assessment of the role of the teacher and of the students' opinion which course modes are most suitable for them, i.e., which modes produce the best results, by a large number of students. Almost 50 % of the respondents state that the communication methods in the e-course are mostly unnecessary or ineffective. Thus, respondents often find discussions more successful in the classroom than in e-course (23 %), it is easier for them to communicate with teacher/assistant in person (23 %), and rarely rate the e-course communication about learning topics, content and activities with other colleagues as better (7 %).

Scores for the attitudes were analyzed in relation to the groups of respondents defined by certain independent variables. Significant differences in rating are evident only in relation to study experience (years of studying) and attitude towards the course, and there is no significant differences according to gender or field of study in none of the statements in this part of the questionnaire. Students with less experience (first year students) rate specific communication channels of the e-course significantly higher than more experienced students; they find e-discussions more successful than the ones done in the classroom (3.0 ± 1.2 vs. 2.5 ± 1.3, ANOVA, $F = 5.1$, $p = 0.007$), they also find communication with other students better (2.9 ± 1.2 vs. 2.3 ± 1.2,

Table 5 Students' Attitudes Towards E-Courses in General

	Statement	Answer / %					Mean	SD
		1	2	3	4	5		
S14	E-course helps me to organize my learning process better.	5	12	27	36	20	3.5	1.1
S15	I achieve better results in the courses which combine online and classroom mode of learning.	8	7	33	31	22	3.5	1.1
S16	In e-courses I participate more actively and complete my assignments more regularly than in exclusively classroom courses.	15	10	32	25	18	3.2	1.3
S17	It is important and useful to have unlimited access to all materials.	2	2	8	22	66	4.5	0.9
S18	E-course learning materials are more suitable for students' needs.	3	7	25	44	22	3.7	1.0
S19	Discussions in e-course are more successful than in the classroom.	23	20	27	19	11	2.8	1.3
S20	Through e-course I communicate better with other students about the learning topics, content and activities.	26	23	26	19	7	2.6	1.2
S21	E-courses provide easier communication with Teacher/assistant than it is in person.	23	17	31	18	11	2.8	1.3
S22	Face-to-face contact with my teacher is fairly important to encourage the learning process.	9	10	25	24	32	3.6	1.3
S23	It bothers me that in e-learning I am distracted by other online activities (video games, social networks, etc.).	18	19	22	22	19	3.1	1.4

Table 6 Students' Attitudes Towards E-Learning in Relation to the Course Mode Preference

| | Blended | | Online | | Classroom | | ANOVA | |
Statement	Mean	SD	Mean	SD	Mean	SD	F	p
S14 E-course helps me to organize my study process better.	3.6	1.1	4.0	1.1	2.8	1.1	13.55	<0.001
S15 I achieve better results in the courses which combine online and classroom mode of learning.	3.6	1.1	4.1	0.9	2.7	1.0	21.97	<0.001
S16 In e-courses I participate more actively and complete my assignments more regularly than in exclusively classroom courses.	3.3	1.2	4.0	1.0	2.3	1.2	21.89	<0.001
S17 It is important and useful to have unlimited access to all the materials.	4.6	0.8	4.7	0.6	3.9	1.0	14.16	<0.001
S18 E-course learning materials are more suitable for students' needs.	3.8	0.9	4.2	0.8	3.2	1.0	14.07	<0.001
S19 Discussions in e-course are more successful than in the classroom.	2.8	1.3	3.7	1.0	2.1	1.1	14.67	<0.001
S20 Through e-course I communicate better with other students about the learning topics, content and activities.	2.6	1.2	3.4	1.3	2.1	1.0	9.52	<0.001
S21 E-courses provide easier communication with teacher / assistant than it is in person.	2.8	1.2	3.8	1.3	2.0	1.1	16.62	<0.001
S22 Face-to-face contact with my teacher is fairly important to encourage the Learning process.	3.6	1.3	2.6	1.2	4.2	1.1	13.97	<0.001
S23 It bothers me that in e-learning I am distracted by other online activities (video games, social networks, listening to music etc.).	3.0	1.4	2.7	1.4	3.5	1.3	3.39	0.035

ANOVA, F = 6,24, p = 0.002), and they think the communication with the teacher is easier (3.0 ± 1.2 vs. 2.6 ± 1.3, ANOVA, F = 3.28, p = 0.039). As in previous analysis, the attitude towards course significantly influences the scores for e-learning attitudes. In all of the statements in this part of the questionnaire there are statistically significant differences in rating in relation to the preferred course mode (Table 6.). With the exception of the statements S22 and S23, which describe potential disruptions in e-learning (the lack of face-to-face contact with the teacher and distraction by other online content), the students who prefer traditional course mode rate the attitudes towards e-leaving remarkably lower than the students preferring blended or fully online course mode.

1.4.5 Working with the MudRi System

Some of the questions in the survey referred to the students' experiences of working with the MudRi system, and their personal assessment of the ease of working online was required, as well as of the system's suitability for the students' needs (Table 7.). A large number of students (over 90 % "mostly correct" and "entirely correct" answers) think they possess enough computer literacy to successfully use the MudRi system. Students mostly like the system's Interface (54 %), and 77 % of the respondents find the navigation easy for the user. The users should be better informed about the ability to contact the MudRi administrator in case of technical problems (less than half of the respondents are aware of this possibility). Majority of the students have no problems accessing the MudRi system (70 % of the respondents never had any problems) nor opening files in the system (76 %).

1.4.6 Factor Analysis: The Actual Existence of E-Learning Elements in the Used E-Courses

Latent dimensions were determined by factor analysis (KMO = 0,824, Bartlett's sphere test is statistically significant at the level of $p < 0.001$) under the component model using GK criterion for limiting factor extraction (higher than 1), basic solution was transformed into orthogonal varimax position. In the final factor model there were 11 variables, three factors were obtained which interpret 57 % of the total variance, the first factor interprets 34 %, the second one 12.72 % and the third one Interprets 10.28 % of the total variance, Two statements (5 and 12, see Table 1.) were left out of the analysis, because they did not meet the requirements of the Thurston principle of the simple structure (variable is measuring the subject properly if the variable

Table 7   Students' Experiences of Working with the MudRi System

	Statement	Answer / %					Mean	SD
		1	2	3	4	5		
SUS 1	My computer literacy is sufficient for the successful use of e-course and working with MudRI system.	1	2	7	13	78	4.7	0.7
SUS 2	Often I have problem accessing MudRi and my e-courses.	50	20	0	20	10	2.2	1.1
SUS 3	Often I have problems opening files on MudRi.	55	21	12	7	5	1.9	1.2
SUS 4	When I encounters a technical problem, I know I can contact the MudRI administrator.	14	13	29	20	24	3.3	1.3
SUS 5	I like the interface of the MudRi system.	7	7	33	35	19	3.5	1.1
SUS 6	The navigation in the MudRi system is intuitive and easy.	2	7	15	37	40	4.1	1.0

saturations in other dimensions are particularly low). The first factor consists of five statements: 1,2,3,11 and 13 (Table 8.) and interprets 34 % of the total variance. The statements refer to the organization, clarity, order, existence of the needed materials as well as to other dimensions of the quality learning material. Furthermore, there are statements referring to the role of the teacher, and since regular editing of the materials and the feedback are also part of this factor, it is called "Learning materials and teacher". The second factor includes three statements: 6, 7 and 8, and interprets 12.7 % of the total variance. The statements refer to the possible and simplified communication with other students and teacher / assistant in e-courses, especially through Forums, therefore this factor is called "Online Communication and activities". The third factor consists of three statements as well: 4, 9 and 10, and interprets 10.3 % of the total variance, The statements refer to the *additional value* in e-courses which requires great preparation, skills and competences from the teachers. It concerns the availability of mandatory and optional study material in digital form, knowledge self-assessment tests and the use of multimedia. This factor is called "E-help in learning" Cronbach reliability coefficient is calculated for every set of variables in a certain factor. In the first factor, the Alpha coefficient was 0.775 (4 variables). In the second factor, the Alpha coefficient was 0.634 (3 variables). In the third factor, the Alpha coefficient was 0.570 (3 variables). In general, we can say that these scales are a reliable set of features. Among the factors there is a statistically significant difference in scores (factor "Help in e-learning" has significantly lower score than the rest). It can be concluded from the rating that the elements of the factor "Help in e-learning" are not very present in the e-courses (self-assessment tests, e-literature and multimedia).

Scores for certain factors were also analyzed according to the groups of students defined by independent variables. Blended E-Learning in Higher Education In relation to gender, there are significant differences only in the factor "Learning materials and teacher", where female respondents give significantly higher score (4.1 ± 0.9 vs. 3.9 ± 0.9, $t=4.81$, $p<0.001$). According to field of study, there is also a significant difference only in the factor "Learning materials and teacher", where SSH students give significantly higher score than the other groups (4.2 ± 0.9 vs. 3.9 ± 0.9 ICT students and 4.0 ± 0.9 ENG-MATSCI students, ANOVA, $F=6.43$, $p=0.001$). Regarding study experience, the differences exist only in the factor "Online communication and activities", where least experienced students give significantly higher scores (4.0 ± 1.2 with first year students, in contrast to 3.6 ± 13 with older students, $F=8.85$, $p<0.001$). Regarding e-learning experience, the difference in scores occurs in two factors "Online communication and activities" and "E-help in learning".

Table 8   The Structure of Latent Dimensions: The Actual Existence of Certain E-learning Elements

Factor Title	Statement	Factor 1	2	3	Mean	SD	Factor Mean	Factor SD
Learning Materials and Teachers	I3	The learning materials and activities in the e-course were well organized.	0.79			4.0	0.9	4.0	0.9
	I2	Learning materials in the e-course were written in a clear and understanding manner, they were delicately colored, and had a simple standardized form.	0.79			4.1	0.9		
	I1	The e-course provided all the materials needed for achieving the expected learning results.	0.75			3.9	1.0		
	I11	E-course teacher edited content and managed e-course activities regularly.	0.56			4.1	0.9		
	I13	I regularly received feedback about my work from e-course teacher.	0.54			3.9	1.0		
Online Communication and Activities	I7	Through the e-course I communicated with other colleagues from the group.		0.74		3.1	1.4	3.8	1.3
	I8	E-course enabled Forum discussions.		0.71		4.4	0.9		
	I6	It was easy to communicate with Teacher/assistant through the e-course.	0.38	0.70		3.8	1.1		
Help in E-Learning	I10	E-course provided mandatory and optional study material in digital form.			0.82	3.0	1.3	2.9	13
	I9	E-course provided ways to test knowledge through self-assessment.			0.71	2.9	1.4		
	I4	Multimedia (appropriate audio and video content, animations, computer simulations, etc.) was used in the e-course.	0.36		0.55	2.9	1.3		

Thus, the inexperienced e-course users (up to one year) give significantly higher scores for the factor "Online communication and activities" (3.9±1.3 vs. 3.6±1.3 with more experienced users, t=2.76, p=0.006), as well as the factor "e-help in learning" (3.0±1.3 vs. 2.8±1.3, t=1.97, p=0.049). The attitude towards courses differentiates respondents according to the rating of the factor "Learning materials and teacher" and "Online communication and activities". Thus, the students who prefer traditional classroom courses give significantly lower scores than others for both factors (for "Learning materials and teacher" 3.7±1.0 vs. 4.0±0.9 online and 4.1±0.9 blended courses, F=18.11, p<0.001, and for "Online communication and activities" 3.2±1.3 vs. 3.8±1.2 online and 4.1±1.3 blended courses, F=18.18, p<0.001). Students rate factor 1 differently according to their grade point average, thus the finest students give the highest scores, while the less good students give significantly lower scores for the factor "Learning materials and teacher" (4.1±0.9 students with grade point average >4.5 vs. 3.8±0.9 students with grade point average < 3.5, F=5.103, p=0.002). Other factors do not illustrate significant differences between students according to the success in their studies.

1.4.7 Factor Analysis: The Importance of the Existence of Certain E-Learning Elements in E-Courses

Latent dimensions were determined by factor analysis (KMO=0,845, and Bartlett's sphere test is statistically significant at the level of p < 0.001) under the component model using GK criterion for limiting factor extraction (higher than 1), basic solution was transformed into orthogonal varimax position. In the final factor model there were 10 variables, two factors were obtained which interpret 61.9 % of the total variance, the first factor interprets 42.2 % and the second one 19.8 % of the total variance. Three statements (4,5 and 12, see Table 2.) were left out of the analysis, because they did not meet the requirements of the Thurston principle of the simple structure. Similar to the instrument used for assessing the existence of certain elements, in this case as well a similar latent structure was obtained. However, the third factor was not obtained which can be interpreted as the absence of the mentioned elements, but can also be explained by the lesser importance that students attribute to certain elements. The first factor involves six statements: 1,2,3,10,11 and 13 (Table 9.) and interprets 42.2 % of the total variance. The statements refer to the perception of importance of the e-course elements such as organization, clarity, order and existence of the required materials, as well as other dimensions of the quality learning material. Furthermore, there are statements referring to the

Table 9　The Structure of Latent Dimensions: The Importance of Certain E-Course Elements

Factor Title	Statement	Factor 1	Factor 2	Mean	SD	Factor Mean	Factor SD
Learning Materials and Teacher	S3	It is important to me that learning materials and activities in the e-course are well organized.	0.83		4.6	0.7	4.6	0.7
	S11	It is important to me that e-course teacher edits content and manages e-course activities regularly.	0.81		4.7	0.6		
	S1	It is important to me that, e-course provides all the materials needed for achieving the expected learning results.	0.79		4.6	0.7		
	S2	It is important to me that learning materials in the e-course are written in a clear and understanding manner, that they are delicately colored, and have a simple standardized form.	0.77		4.5	0.7		
	S13	It is important to me that I regularly receive feedback about my work from e-course teacher.	0.68		4.6	0.7		
	S10	It is important to me that e-course provides mandatory and optional study material in digital form.	0.57	0.36	4.4	0.9		
Online Communication and Activities	S7	It is important to me that through the e-course I communicate with other colleagues from the group.		0.90	3.4	1.3	3.7	1.2
	S8	It is important to me that e-course enables Forum discussions.		0.86	3.6	1.2		
	S6	It is important to me that it is easy to communicate with teacher/assistant through the e-course.		0.82	3.8	1.1		
	S9	It is important to me that e-course provides ways to test knowledge through self-assessment.	0.35	0.52	3.9	1.1		

role of the teacher, and since regular editing of the materials and the feedback are also part of this factor, it is called "Learning materials and teacher". The second factor consists of four statements: 6,7,8 and 9, and interprets 19.8 % of the total variance. These statements refer to the importance students attribute to the e-course elements such as communication with other students and teacher/assistant, as well as communication via Forum, but also to the existence of the knowledge self-assessment tests. This factor is called "Online communication and activities". Cronbach reliability coefficient is calculated for every set of variables in a certain factor. In the first factor, the Alpha coefficient was 0.842 (6 variables). In the second factor, the Alpha coefficient was 0.814 (4 variables).

When assessing importance, the factor "Learning materials and teacher" is rated significantly higher than the actual existence of corresponding elements $(4.6 \pm 0.7$ vs. 4.0 ± 0.9, comparison with the data in Table 8.), while the factor "Online communication and activities" is assessed with the same score that is present for the existence of the corresponding elements $(3.7 \pm 1.2$ vs. 3.8 ± 1.3, comparison with the data in Table 8.).

The scores for the factor of importance of certain e-learning elements were also analyzed according to the groups of students defined by independent variables. Regarding gender, there are significant differences in rating of the factor "Learning materials and teacher" where female respondents give significantly higher scores $(4.7 \pm 0.6$ vs. $4.5 \pm 0.7.1 = 4.29$, $p < 0.001$). According to field of study, there is a significant difference only in the first factor "Learning materials and teacher" where SSH students give if significantly higher importance than other groups of students $(4.7 \pm 0.6$ vs. 4.6 ± 0.6 ICT students and 4.5 ± 0.7 ENGMATSCI students, $F = 10.37$, $p < 0.001$). Regarding study experience, the difference exists only in the factor "Online communication and activities", where least experienced students give significantly higher importance to these e-learning elements (score 3.9 ± 1.1 with first year students, in contrast to 3.6 ± 1.2 with older students, $F = 11.12$, $p < 0.001$). According to experience with e-courses, there are differences in rating of both factors. The least experienced e-course users (up to one year) give significantly higher importance to the factor "Learning materials and teacher" $(4.6 \pm 0.7$ vs. 4.4 ± 0.6 with experienced users, $t = 2.01$, $p = 0.049$) as well as to the factor "Online communication and activities" $(3.8 \pm 1.1$ vs. 3.6 ± 1.2, $t = 3.31$, $p < 0.001$).

Regarding attitude towards courses, the respondents assess the importance of existence of certain e-learning elements differently; thus, the respondents who prefer traditional classroom courses give significantly lower importance to the factor "Learning materials and teacher" (4.4±0.8 vs. 4.8±0.6 on-line and 4.7±0.6 blended courses, F=17.23, p<0.001), as well as to the factor "Online communication and activities" (3.1±1.1 vs. 4.2±1.0 online and 3.8±1.2 blended courses, F=39.38, p<0.001). According to students" grade point average, there are differences in the factor "Learning materials and teacher", which better students value more than the less good students (4.6±0.6 better students vs. 4.4±0.9 students with grade point average <2.5, F=4.18, p=0.005).

1.4.8 Factor Analysis: Attitudes towards E-Learning

Latent dimensions were determined by factor analysis (KMO=0,847 Bartlett's sphere test is statistically significant at the level of p < 0.001) under the component model using GK criterion for limiting factor extraction (higher than 1), basic solution was transformed into orthogonal varimax position. All of the variables were included in the final factor model. Three factors were obtained which interpret 68 % of the total variance, the first factor interprets 42.1 %, the second one 14.1 %, and the third one 11.8 % of the total variance. The first factor consists of five statements: 14,15,16,17 and 18 (Table 10.) and interprets 42.1 % of the total variance. The statements refer to the specific features of blended courses, such as achieving better results in the courses which combine blended mode, better organization of learning and better suitability of the materials which are available at any time. Furthermore, students prefer e-courses because they become more active and complete their assignments more regularly than in the traditional classroom. This factor is called "Specific features of blended courses". The second factor includes three statements; 19, 20 and 21, and interprets 14.1 % of the total variance. The statements refer to the perception of better, easier and more effective communication, both with other students and the teacher/assistant. This factor is called "Online communication". The third factor interprets 11.8 % of the total variance and consists of only two statements (22 and 23), Internet is perceived as the source of distraction and loss of concentration due to the abundance of content that it offers. Furthermore, the statements refers to the importance of face-to-face contact to encourage learning process, therefore this factor is called "E-disruptors". Cronbach reliability coefficient is calculated for every set of variables in a certain factor. In the first factor, the Alpha coefficient was 0.836 (5 variables). In the second factor, the Alpha coefficient was 0.854 (3 variables). In the third factor, the Alpha coefficient was 0.343 (2 variables), therefore, in future analysis additional variables

Table 10 Structure of Latent Dimensions: General Attitudes Towards E-Learning

Factor Title	Statement	Factor			Mean	SD	Factor Mean	Factor SD
		1	2	3				
Specific Features of Blended Courses	S15 I achieve better results in the courses which combine online and classroom mode of learning.	0.78		0.27	3.5	1.1	3.7	1.2
	S14 E-course helps me to organize my learning process better.	0.77		0.25	3.5	1.1		
	S18 E-course learning materials are more suitable for students' needs.	0.73			3.7	1.0		
	S17 It is important and useful to have unlimited access to all the materials.	0.70			4.5	0.9		
	S16 In e-courses I participate more actively and complete my assignments more regularly than in exclusively classroom courses.	0.69		0.40	3.2	1.3		
Online Communication	S20 Through e-course I communicate better with other students about the learning topics, content and activities.		0.85		2.6	1.2	2.7	1.3
	S21 E-courses provide easier communication with teacher/assistant than it is in person.		0.83		2.8	1.3		
	S19 Discussions in e-course are more successful than in the classroom.	0.28	0.81		2.8	1.3		
E-Disruptions	S23 It bothers me that in e-learning I am distracted by other online activities (video games, social networks, listening to music, etc.).			0.84	3.1	1.4	3.3	1.3
	S22 Face-to-face contact with my teacher is fairly important to encourage the learning process.		−0.42	0.64	3.6	1.3		

should be added to this factor. Scores for certain factors in general attitudes towards e-learning were also analyzed according to the groups of students defined by Independent variables. Regarding gender, there are no significant differences in attitudes in any of the factors. According to Field of study, there is a significant difference only in the first factor, where the factor "Specific features of blended courses" receive significantly higher score from ICT students than from other respondents (3.8 ± 1.1 vs. 3.6 ± 1.2 with all other students, $F=3.39$, $p=0.034$). Regarding study experience, differences occur in the factor "Specific features of blended courses" and the factor "Online communication", where least experienced students give significantly higher score for these elements (score 3.8 ± 1.1 with first year students, in contrast to 3.6 ± 1.2 with older students, $F=11.12$, $p<0.001$ for the first factor and score 3.0 ± 1.2 with first year students, in contrast to 2.5 ± 1.3 with older students, $F=14.21$, $p<0.001$ for the second factor). According to e-learning experience, the differences in rating occur only in the factor "Online communication", where least experienced e-course users (up to one year) give significantly higher scores (2.8 ± 1.3 vs. 2.6 ± 1.3 with experienced users, $t=2.77$, $p=0.006$). Regarding attitude towards the course, respondents express significantly different attitudes towards certain e-learning elements; thus, the students who prefer traditional classroom courses give lower scores for the factor "Specific features of blended courses" (3.0 ± 1.2 vs. 4.2 ± 0.9 online and 3.8 ± 1.1, blended courses, $F=71.71$, $p<0.001$) and the factor "Online communication" (2.1 ± 1.1 vs. 3.6 ± 1.2 online and 2.8 ± 1.3 blended courses, $F=40.26$, $p<0.001$), while the situation is completely opposite with the factor "E-disruptors" - the respondents who prefer traditional classroom courses give significantly higher scores than the others (3.8 ± 1.3 vs. 2.6 ± 1.3 online and 3.3 ± 1.3 blended courses, $F=13.89$, $p<0.001$). According to students' success in their studies (grade point average), there are no differences in the assessment of the attitudes towards e-learning.

1.5 Conclusion

As the e-learning support and platform is active in University of Rijeka only for two years, the primary objective of this study was to analyze the current state of student's perception and acceptance of e-learning as a new educational tool. Secondly, the aim of the study was to get the student's feedback on the value and importance of the specific elements of e-learning implemented in e-courses. Finally, the same study aimed at getting to know the student's general attitudes towards e-learning and detect their needs in blended courses.

The data collected through this research and obtained results were meant to serve as a feedback to all instances supporting e-learning implementation, to design guidelines for the teachers regarding student's needs and preferences in e-courses and to provide data on users' profiles. The research was designed as a cross sectional study in which the data were collected through an online survey distributed to all students using e-learning platform at the time. The questionnaire was developed solely for the purpose of this study, but using the guidelines from similar researches on the acceptance of information technology and users' satisfaction (Bernard et al., 2004; Davis, 1989; DeLone & McLean, 2003; Poelmans, 2009).

Items were adapted to suit our research questions, namely to gather the data on the student's perception of quality of already delivered e-courses, to find out the level of importance for the specific elements of e-learning and to get to know student's general attitude toward e-learning as well as their needs with respect to quality of course materials, communication and support of the learning process. The response rate was satisfactory and acquired sample representative of our students' population using e-learning tools with respect to socio-demographic characteristics and student's profiles. Participants in the study asses the current state of e-learning elements implementation quite good; they agree the educational materials are in most cases complete, organized and well designed, and that teachers engage in online work well; they perceive the teachers manage the e-courses well, communicate regularly and timely provide the feedback. The lower level of agreement is obtained on the use of multimedia, offering of the self-assessment tests, accessibility of digital literature and collaborative activities. This suggests teachers should be encouraged (and trained) to put more effort in designing and offering suitable multimedia elements to enrich their materials, self-assessment test to make students feel more comfortable in terms of examination expectations, and to design online activities for the students to enhance collaborative aspects in teaching. As the working with MudRi LMS is concerned, learners consider themselves sufficiently IT competent to successfully use e-courses and work with the system, and the majority agrees that MudRi is user friendly and has nice interface. Importantly, majority never have had problems with accessing nor with opening files on system. However, students are not well informed about the possibility to ask for technical assistance, suggesting that e-learning support service should work, on informing users (particularly students) better. The results from "general importance of specific e-learning elements" part of the survey indicate that students value the most the completeness, organization and design of educational materials, as well as teacher's online engagement, especially in good management of e-course, in regular communication and timely providing feedback. They do not perceive as much as important the online activities, communication to other students and discussions. When the comparison of "current state" and "general importance" for the specific e-learning elements is made, it seems that there is not much of discrepancy; the level of agreement with perceived state of implementation and wished state of implementation, is for all the elements just a small portion below (on average for 0, 5, on the scale of 1–5). Assessing the general value of e-learning and its characteristic, students best agree with the notion that most important is to have the access to teaching materials 24/7. Second best is that online materials are better suited to students' needs and that in general, having e-course as addition to classroom teaching is helping them organizing their learning better and overall achieving better results. They do not comply with statements of online discussions being more successful than classroom discussions and online communication to teachers and colleagues being better. Generally, the majority point out that f2f communication to teachers is important in facilitating the learning process. As the general attitude towards online learning is considered, it is interesting that preferences for exclusively online and/or blended learning are dominant in a group of students having better average studying grades (A or B), as well as in a group having shorter studying experience (1–2 ys of studying). The same groups find the new communication channels (online discussion forums, mailing with teachers, assistants and colleagues within the online learning environment) important and useful. Moreover, they think that online educational materials and activities are better suited to students' needs and that they help them achieve the learning outcomes better. This results is an important signal to all instances supporting e-learning implementation (from university management to supporting services), since it suggests that the "most wanted" users (students with better grades and fresh students) are those that willingly accept technology in learning. However, irrespective of their learning preferences, the majority of student judge online discussions and online communication to teachers and colleagues to be not very valuable; this result points towards further investigation of quality of implementation of collaborative learning activities. Most probably this attitude arises from student's lack of good experience with online collaboration, suggesting the need to enhance teachers' competencies for online teaching, particularly in acquiring successful tutoring methods and learners' support methods (MacDonald, 2008; Wilson, 2004). We think that continuous and careful monitoring of learner's satisfaction will ensure the success, feasibility and viability of online learning, as supporting educational tool in university study programs. Blended learning systems change the way the learners learn (Graham, 2006), but also change the way the teachers teach. This process of transformation cannot happen overnight and is expected to last for some time, but we hope that it will bring alongside also some quality changes in organization, planning and management of higher education to our University, which can in turn bring about higher quality of education.

References

1 Bates, A. W. (1999). Managing technological change: Strategies for academic leaders. San Francisco: Jossey Bass.

2 Bernard, R. ML, Brauer, A., Abrami, P. C, & Surkes, M. (2004). The development of a questionnaire for predicting online learning achievement Distance Education, 25(1), 31–47, Retrieved January 12, 2009, from the http ://www.ebscohost.com/

3 Bonk, C. J., (2009), The world is open: How web technology is revolutionizing education. Jossey-Bass.

4 Davis, F. D. (1989), Perceived usefulness, perceived ease of use, and user acceptance of information technology. MIS Quarterly, 13(3), 319–340,

5 DeLone, W., & McLean, E. R. (2003). The DeLone and McLean model of information systems success: A ten-year update. Journal of Management Information Systems, 19(4), 9–30.

6 Duderstadt, J. J., Atkins, D, E., & Van Houweling, D. (2003), The development of institutional strategies Educause Review, 38(3), 48–58.

7 Ehlers U-D. (2004), Quality in e-learning from a learner's perspective. European Journal of Open, Distance and E-Learning, Retrieved September 12, 2010, from http://www.eurodl.org/index.php?article=l0l

8 Ellis, R. A., Jarkey, N., Mahony, M, J., Peat, M. & Sheely, S. (2007). Managing quality improvement of e-Learning in large, campus-based university. Quality in Higher Education, 15(1), 9–23.

9 Graham, C. R. (2006). Blended learning systems. In C. J. Bonk & C. R. Graham, The handbook of blended learning; Global perspectives, I ocal designs. Pfeiffer.

10 Hanna, E. D. (2003). Building a leadership vision - Eleven strategic challenges for higher education. Educause Review Magazine, 38(4), 25–34.

11 Keller, C., & Cernerud, L. (2002). Students' perceptions of e-learning in university education. Journal of Educational Media, 27. Retrieved June 20, 2006 from http://www tandf.co.uk/journals/titles/13581651.asp.

12 MacDonald, J. (2008). Blended learning and online tutoring: Planning learner support and activity design. Gower Publishing Company.

13 McCradie, J. (2003), Does IT matter to higher education? Educause Review, 38(6), 15–22.

14 Poelmans, S., Wessa, P., Milis, K,Bloemen, E., & Doom, C. (2009). Usability and acceptance of e-learnling In statistics education, based on the compendium platform. Retrieved May 12, 2010 from http://www.wessa.net/download/iceripaper1.pdf

15 Roncevic, N.t Ledic, J. & Vrcelj, S. (2009), The predictive validity of the instrument and learning outcomes: A case study of hybrid teaching at the University of Rijeka, In D. Komlenovic (Ed.), International Conference Quality and Efficiency of Teaching In Learning Society. Book of abstracts Beograd: Institut za pedagoska istrazivanja, p. 251–251.

16 Wagner, G. D., & Flannery, D. D. (2004). A quantitative study of factors affecting learner acceptance of a computer-based training support tool Journal of European Industrial Training, 28(5), 383–399.

17 Wilson, G. (2004). Online interaction impacts on learning: Teaching the teachers to teach online. Australasian Journal of Educational Technology, 20(1), 33–48.

18 Zuvic-Butorac, M., & Nebic, Z. (2009). Institutional support for e-learning implementation in higher education practice; A case report of University of Rijeka, Croatia. Proceedings of the 'international Conference on Information Technology Interfaces, ITI, art. no. 5196130, 479–484.

2

The Impact of Information Systems on the Performance of Human Resources Department

Usman Sadiq[1], Ahmad Fareed Khan[1], Khurram Ikhlaq[1]
and Bahaudin G. Mujtaba[2]

[1]*Superior University, Pakistan*
[2]*Nova Southeastern University, Pakistan*

2.1 Abstract

Owing to the revolution in information technology, the face of the contemporary workplace has changed and systems have been made more effective by introducing new techniques. Majority of the organizations have now understood the importance of information storage and retrieval. In this paper, we focus on how modern technology is helping in ensuring effectiveness of HR functions. Human Resource Information System (HRIS) is an opportunity for organizations to make the HR department administratively and strategically participative in operating the organization. The main objective is to understand the extent to which HRIS is being used in increasing the administrative and strategic functions of the HR department. For this purpose, we have conducted a survey of 18 HR Managers from various private corporations operating in Lahore, Pakistan. The results show that HRIS is positively used as a tool to achieve greater administrative efficiency by adding value in the department. However, all of its benefits are difficult o quantify. HRIS utility as a strategic tool is still not been fully recognized, and this is preventing the system to be used to its fullest potential. Suggestions and recommendations are provided.

Keywords: Human resources, Human Resource Information System (HRIS), Private sector banks, Lahore, Pakistan.

2.2 Introduction

The addition of information technology to the human resource industry has revolutionized the contemporary workplace. HR professionals now have an increased capacity not only to gather information, but also to store and retrieve it in a timely and effective manner. This has not only increased the efficiency of the organization but also the effectiveness of management functions. New technology has also created opportunities for higher levels of stress for younger and older workers alike (Mujtaba, Afza, and Habib, N. (2011), unethical temptations and behaviors (Mujtaba, 2011), and opportunities for better leadership practices (Mujtaba and Afza, 2011). The twenty-first century is characterized as the knowledge century (Chin-Loy and Mujtaba, 2007). Most of the organizations are now dependent upon knowledge workers and thus on effective knowledge management practices. Today, knowledge management offers a unique concept considered by many in the industry as progressive and "soft" in applications, primarily because of the intangible elements of knowledge (Mujtaba, 2007, p. 201). The ability to not only attract and hire but also to retain and properly utilize these individuals is crucial knowledge for the survival and success of the organization. In this globalized world, a department that is increasingly becoming central to the implementation of organization policy is the HR department. So the HRIS is now considered an integral part of every organization (Waytt, 2002). More and more organizations are now developing information technology which can help the organization achieve its goals in a timely manner. These information systems can then help the organization make more strategic decisions. HRIS is an effective tool that can be used for streamlining the administrative functions of the HR department. This can be achieved by creating an elaborate and relevant database. The data that an effective HRIS would have on individual employees can include training completed, awards received, projects participated in and finished successfully, level of education attained, number of years of service, skills, competencies, etc.

By using this data the HR department can make a contribution towards strategy formation within an organization. With an efficient HRIS in place, the development of HR systems becomes easier (Dessler, Griffiths, and Walker, 2004).

2.3 Human Resource Management (HRM)

The history of HRM is said to have started in England in early 1800s during the craftsmen and apprenticeship era, and then further developed with the arrival of the Industrial Revolution in the late 1800s. In the 19th century, Frederick W.

Taylor suggested that a combination of scientific management and industrial psychology of workers should be introduced. In this case, it was proposed that workers should be managed not only for the job and its efficiencies but also for the psychology and maximum well-being of the workers. Moreover, with the drastic changes in technology, the growth of organizations, the rise of unions and government's concern and interventions resulted in the development of personnel departments in the 1920s. At this point, personnel administrators were called 'welfare secretaries' (Ivancevich, 2007). HRM is said to have started from the term 'Personnel Management' (PM). The term 'PM' emerges after the Second World War in 1945 as an approach by personnel practitioners to separate and distinguish themselves from other managerial functions and make the personnel function into a professional managerial positions. Traditionally, the function of PM is claimed to 'hire and fire' employees in organizations other than salary payments and training. But there were many criticisms and concerns of ambiguity expressed about the purpose and role of PM to HRM (Tyson, 1985). Therefore, the term HRM gradually tended to replace the term PM (Lloyd and Rawlinson, 1992). However, writers argue that the term HRM has no appreciable difference from PM as they are both concerned with the functions of obtaining, organizing, and motivating human resources required by organizations. At the same time, writers are defining the terms HRM and PM in many different ways (Beer and Spector, 1985), The rebranding of the term from PM to HRM was done due to the evolvement and changes in the world of management and therefore, a contemporary term would seem appropriate that can encompass new ideas, concepts and philosophies of human resources (Noon, 1992, Armstrong, 2000), Indeed, some writers comment that there are 'little differences' between PM and HRM and it has been criticized as pouring 'old wine into new bottle' with a different label (Legge, 2005). Whether HRM is considered to be different than personnel management is a continued debate on both its meaning and practices (Marchington & Wilkinson, 2002; Legge, 2005). Strategic Human Resource Management (SHRM) has grown considerably in the last years. Schuler, Dolan and Jackson (2001) described the evolution of SHRM from personnel management in terms of a two-phased transformation: first from personnel management to traditional human resource management (THRM), and then from THRM to SHRM. To improve performance and create a competitive advantage, a firm's HR must focus on a new set of priorities. These new priorities are more strategic oriented and less geared towards traditional HR functions such as staffing, training, appraisal and compensation. Strategic priorities include team-based job designs, flexible workforces, quality improvement practices, employee empowerment, and incentive-based compensation. SHRM is designed to diagnose strategic needs and plan talent development, which are required to implement a competitive strategy and achieve operational goals (Huselid, Jackson and Schuler, 1997).

2.4 Human Resource Information System (HRIS)

HRIS has a very humble historical origin. Although there were some exceptions, prior to World War II HR professionals (then referred to as "personnel" staff) performed basic employee record keeping as a service function with limited interaction on core business mission. Initial efforts to manage information about personnel were frequently limited to employee names and addresses, and perhaps some employment history often scribbled on 3x5 note cards (Kavanaugh, Gueutal and Tannenbaum, 1990). Between 1945 and 1960, organizations became more aware of human capital issues and began to develop formal processes for selection and development of employees. At the same time, organizations began to recognize the importance of employees' morale on the firm's overall effectiveness. While this period of change in the profession did not result in significant changes in HRIS (although employee files did become somewhat more complex), some believe that it set the stage for an explosion of changes that began in the 1960s and 1970s (Kavanaugh, Gueutal and Tannenbaum, 1990).

During the next twenty years (1960 to 1980) HR was integrated into the core business mission and, at the same time period, governmental and regulatory reporting requirements for employees also increased significantly. The advent and widespread use of mainframe computers in corporate America corresponded with this regulatory increase and provided a technological solution to the increased analytical and record-keeping requirements imposed by growing regulation of employment and a host of new reporting requirements (e.g., affirmative action, EEO, OSHA, etc.). The Human Resource Department became one of the most important users of the costly computing systems of the day, often edging other functional areas for computer access. Although HRIS systems were computerized and grew extensively in size and scope during this period, they remained (for the most part) simple record-keeping systems (Kavanaugh et al., 1990). According to Kovach and colleagues, HRIS is considered as a systematic procedure for collecting, storing, maintaining, and recovering data required by an organization about their human resources, personnel activities and organizational characteristics (Kovach, Hughes, Pagan and Maggitti, 2002).

2.5 Benefits of Human Resource Information System

The rationale for the implementation of HRIS varies between organizations. Some use it to reduce costs, others to facilitate better communication, and some use it to re-orient HR operations to increase the department's strategic contribution (Parry, Tyson, Selbie, & Leighton 2007). HRIS provides management with strategic data not only in recruitment and retention strategies, but also in merging HRIS data into large-scale corporate strategy. The data collected from HRIS provides management with decision-making tool. An HRIS can have a wide range of usage from simple spread sheets to complex calculations performed easily (Parry 2010). Through proper HR management, firms are able to perform calculations that have effects on the business as a whole. Such calculations include health-care costs per employee, pay benefits as a percentage of operating expense, cost per hire, return on training, turnover rates and costs, time required to fill certain jobs, return on human capital invested, and human value added. It must be noted that none of these calculations results in cost reduction in the HR function (DeSanctis, 1986: 15). The aforementioned areas, however, may realize significant savings using more complete and current data that can be made available to the appropriate decision makers. Consequently, HRIS is seen to facilitate the provision of quality information to management for informed decision-making. Most notably, it supports the provision of executive reports and summaries for senior management and is crucial for learning organizations that see their human resources as providing a major competitive advantage. HRIS is therefore, a medium that helps HR professionals perform their job roles more effectively (Grallagher, 1986; Broderick and Boudreau, 1992). HRIS can be implemented at three different levels, i.e. the publishing of information, the automation of transaction, and finally transforming the entire working of the HR department so it plays a more strategic role and adds more value to the organization (Lengnick-Hall and Moritz 2003). It is, however, very difficult to ascertain the value addition made by HRIS on the revenues and profits of an organization since strategic HRIS is beneficial in facilitating the decision-making process. These decisions can result in greater employee motivation and satisfaction and both are extremely difficult to quantify (Kovach, Hughes, Fagan, Maggitti, 2002). Mayfield and Lunce (2003) came to a similar conclusion that while administrative activities can be quantified and measured such as reduction in turnover and efficiency of HR department, it is very difficult to attribute certain gains such as motivation and morale directly to the implementation of HRIS. As opposed to administrative HRIS, it is complex to establish a definitive link between organization benefits and HRIS deployment.

The literature shows many previous related studies in HRIS, however, most of them were theoretical (Ngai and Wat, 2006). In addition, most studies were conducted in the context of developed countries' organizations. Ngai and Wat (2006) conducted a survey of the implementation of HRIS in Hong Kong organizations. They found that the greatest benefits of the implementation of HRIS were the quick response and access to information that it brought, while the greatest barrier was the insufficient financial support. In addition, Ngai and Wat (2006) reported many other previous related studies conducted in HRIS implementation. For example, a study of Martinson's (1994) aimed to compare the degree and sophistication in the use of IT between Canada and Hong Kong. Martinson found that the use of HRIS was less widespread in Hong Kong than in Canada, while IT for HRM was applied more in Hong Kong than in Canada. Ball (2001) conducted a survey in order to explore the uses of HRIS in smaller UK organizations and found that smaller organizations were less likely to use HRIS.

Moreover, Burbach and Dundon (2005) conducted a study to assess the strategic potential of HRIS to facilitate people management activities in 520 organizations in the Republic of Ireland, They found that foreign owned large organizations adopted HRIS more often than smaller Irish owned organizations. They also found that HRIS technologies were used for administrative rather than strategic decision-making purposes. Another recent study conducted by Delorme and Arcand (2010), aimed to elaborate on the development of the roles and responsibilities of HR practitioners from a traditional perspective to a strategic perspective, found that the introduction of new technologies in the organization affected the way HR professionals accomplished their tasks within the HR department and the rest of the organization. The study of Krishnan and Singh (2006) explored the issues and barriers faced by nine Indian organizations in implementing and managing HRIS. The main HRIS problems were lack of knowledge of HR department about HRIS and lack of importance given to HR department in these organizations. Cooperation is required across various functions and divisions of the organization for proper implementation of HRIS.

2.6 Methodology and Results

The data used in this research is qualitative and specifically gathered by the authors for this study. A survey was developed and given to the Human

Resource Directors of private banks working in Lahore (Punjab), Pakistan. The objective was to assess the administrative and strategic impact of HRIS in Pakistan. Pakistan is a country going through developments, opportunities and challenges like any other nation in today's twenty-first century competitive global workplace (Yasmeen, Begum, and Mujtaba, 2011). As such, this study of human resources professionals and the implementation of new technology can be a good initiative towards efficiency and productivity. A Likert-type items on a five point scale and open-ended questions were employed on the survey to measure the perceptions of the HR directors in regard to the impact of the HRIS on HR processes, the time spent on various HR activities, the expense of HR activities, levels and use of information within the organization, the role of the HR department, and strategic decision making. From the 20 surveys given to HR professionals, eighteen completed and usable surveys were used for drawing conclusions in this study. The questionnaire used in this survey was adopted from a study completed by Beadles, Jones and lowery (2005). There is a considerable gap in previous researches when it conies to analyzing the impact of HRIS on Pakistan's banking sector. This study aims to bridge some of that gap.

This research was exploratory and primarily descriptive in nature. The intent was to discover whether HR directors perceived that human resource information systems were proving beneficial in regard to its strategic impact on the organization. The survey items are contained in Figures, The results of the survey are contained in Tables (1–7). The survey items were divided into categories concerning satisfaction with the HRIS (Table 1); the impact of the HRIS on HR processes (Table 2); time savings due to the HRIS (Table 3); the effect of the HRIS on expenses (Table 4), information effects (Table 5), decision-making (Table 6); and the strategic impact of the HRIS and the impact of the HRIS on the role of the HR function in the organization (Table 7). We had a relatively small sample size as mentioned above. Therefore, we used frequency tables to measure the percentage of favorable responses to a series

Table 1 Satisfaction with HRIS

Items	% Agreed
Overall I am satisfied with our HRIS.	37.5
The employees of HR department appear to be satisfied with our HRIS.	50.0
Our HRIS has met our expectations.	50.0
Our HRIS could be better utilized.	100.0

Table 2   HR Process

Items	% Agreed
Our HRIS has improved the recruitment process.	75.0
Our HRIS has improved the training process.	50.0
Our HRIS has improved the data input process.	75.0
Our HRIS has improved the data maintenance process.	87.5
Our HRIS has helped with forecasting staffing needs.	87.5
Our HRIS has decreased paper work.	87.5

Table 3   Time Savings

Items	% Agreed
Our HRIS has decreased the time spent on recruiting.	75.0
Our HRIS has decreased the time spent on training.	37.5
Our HRIS has decreased the time spent on making staff decisions.	75.0
Our HRIS has decreased the time spent on inputting data.	62.5
Our HRIS has decreased the time spent on communicating information within our institution.	50.0
Our HRIS has decreased the time spent on processing paper work.	75.0
Our HRIS has decreased the time spent on correcting errors.	62.5

Table 4   Cost Savings

Items	% Agreed
Our HRIS has decreased cost per hire.	37.5
Our HRIS has decreased training expenses.	12.5
Our HRIS has decreased recruiting expenses.	37.5
Our HRIS has decreased data input expense.	62.5
Our HRIS has decreased the overall HR staff's salary expense.	37.5

Table 5   Information Effects

Items	% Agreed
Our HRIS has improved our ability to disseminate information.	37.5
Our HRIS has provided increased levels of useful information.	75.0
The information generated from our HRIS is shared with top administrators.	87.5
The information generated from our HRIS is underutilized by top administrators.	50.0
The information generated from our HRIS has increased coordination between HR department and top administrators.	62.5
The information generated from our HRIS has added value to the institution.	87.5

Table 6   Decision-Making

Items	% Agreed
Our HRIS has made our HR decision-making more effective.	37.5
The information generated from our HRIS helps our institution decide on employee raises.	37.5
The information generated from our HRIS helps our institution to make more effective promotion decisions.	25.0
The information generated from our HRIS helps our institution decide when to hire.	25.0
The information generated from our HRIS helps our institution make better decisions in choosing better people.	37.5
The information generated from our HRIS helps our institution decide when training and skill development are necessary.	37.5

Table 7   Strategic Impact and Role of HR

Items	% Agreed
Our HRIS has made the HR department more important to the institution.	87.5
Overall our administration thinks that HRIS is effective in meeting strategic goals.	37.5
The information generated from our HRIS has improved the strategic decision making of top administrators.	62.5
The information generated from our HRIS has made HR a more strategic partner in the institution.	87.5
Our HRIS has promoted our institution's competitive advantage.	37.5

of questions assessing HR directors' perceptions of HRIS. The expressed results are the percentage of respondents for each item who either agreed or strongly agreed with the statement. Table 1 displays the satisfaction of management with the HRIS system. The results show that, only about half were satisfied with the HRIS, and a similar number of respondents concurred that the system was up to their expectation. The percentage of people actually being satisfied was just above one-third, and all employees agreed that their HRIS could be put to better use. These studies indicated that the satisfaction with regards to HRIS was mixed. And almost all of the respondents felt that it could be better utilized. These results do not take into consideration whether the staff had been trained properly in the use of HRIS, nor were they properly briefed about the systems utility.

With regard to HRIS contribution in streamlining various HR processes, nearly 88% respondents agreed that administrative processes such as decrease in paperwork, forecasting staffing need and data maintenance had indeed improved. Furthermore, 75% of the employees surveyed agreed that data input and recruitment process had been made more efficient. On the question of HRIS having a considerable impact on the training process, the response in favor of it was 50%.

Time saving is one of the barometers against which the efficiency of any IS system can be gauged. This study showed that 75% of the respondents believed that the system had a positive impact on some administrative functions such as time spent on recruiting, routine staff decisions, processing of paper work, and error correction. However, only half believed that it had actually helped in improving the communication of information within the organization. Only a third of the surveyed employees believed that HRIS decreased the time spent on training.

When it came to the actual cost saving from the HRIS in the organization, the results were pretty similar to previous researches, such as the one carried out by Beadles (2005), Only 37.5% of the respondents believed that the HRIS had actually decreased the cost of hiring, the recruitment expenses or the salary of HR staff. Even a lower percentage (12.5%) of respondents thought that training expenses were reduced; however, 62.5% of the respondents believed that administrative tasks such as data input expenses did come down.

The adequate storage and timely retrieval of information is a hallmark of an effective IS system. 75% of the respondents believed that HRIS indeed provided useful information, while a greater number (87.5%) of the respondents believed that the information received through HRIS added value. Whereas an identical percentage (87.5%) felt that generated information was being shared with the top management and only half (50%) believed that this information was actually being utilized by the administrator. These results indicate a lack of willingness to use the information as a strategic tool.

With regard to HRIS helping management in making better decisions, the findings support the results of Beadles (2005) that HRIS is not considered a decision-making instrument. Only a third of the total respondents believed that HRIS contributed to making decisions more effective, and an equal numbered believed that HRIS played a significant role in the selection of better candidates or improving training and development of the staff. Even a lower percentage (25%) said that hiring decisions were made using information available through HRIS. This would indicate that HRIS was viewed rather, favorably as an administrative tool, but not a strategic one.

In terms of whether HRIS has enhanced the strategic role of the HR department, 87.5% respondents believed that HRIS increased the importance of HR department and made it a strategic partner, whereas only 37.5% believed that

HRIS gave a competitive edge to the institution or was effective in helping the organization meet its strategic goals. However, 62.5% did believe that HRIS improved the strategic decision-making of the top administrators.

2.7 Recommendations

In terms of limitations, it should be noted that local banks were included in the survey. The research was limited to Pakistan's banking sector only. The data was only collected from managers. Professional staff members in lower ranks could also be included in future studies as they might have a different view on the use and benefits of HRIS. The number of foreign banks operating in Pakistan is not very high and therefore their input was not included in this study. Other sectors such as textile, manufacturing and private academic institutions also use HRIS to a varying degree, and they can be part of future research studies.

2.8 Impact of HRIS on Hiring

Hiring is usually the last and final step of the recruitment process. Recruitment is one of the most important and fundamental functions of the HR department. An effective recruitment strategy can lead to the hiring of the best candidate. This in turn can contribute not only in keeping cost down, but also in facilitating the processes of succession planning, employee retention, greater employee motivation, and reduced turnover. This is, however, contingent on the HR department having complete information about the nature, demands and construction of the job on one hand but also the knowledge about the personal competencies that are required to fulfill those jobs on the other. The survey showed that only 37.5% of the respondents believed that HRIS played a role in finding suitable candidates, while an equal number believed that HIRS actually brought down the cost of recruitment and the cost per hire. Even a lower number (25%) believed that HRIS was instrumental in deciding when to hire. The reason for these low statistics, when it comes to hiring the right people, can be structural (the size of the organizations), as well as cultural (accepted behaviors within an organization) or simply representative of ground realities (the external environment within which the organization operates). With inadequate background checks and lack of proper references, employers tend to hire people through informal networks of personal Contacts. In this scenario, the HRIS system is of little importance and any real relevance. On the administrative side regarding choosing the best people, the problem might be a lack of user preference in the context of using the HRIS, but greater emphasis would be on how the organization is operated. It can be that, top management simply regards the HR department as merely an administrative tool rather than a serious participant in setting the strategic priorities of the organization. On the HRIS side, if profiles are not properly made and maintained then the selection of the best candidate is difficult. A proper employee's profile should consist of number of years with the company, projects participated in, training attended, certifications completed, awards won, and targets achieved. Future employee aspirations, goals and milestones need to be put in the system. The system must also support a proper succession plan, which would indicate the positions to be vacated, and the basic criteria against which a potential successor would be evaluated. When the profiles of employees can be linked to the succession planning tree, then this can facilitate at least internal hiring, at the right place with minimal cost.

2.9 Role of HRIS in Improving Training

One of the lowest percentages in the survey was the attitude of respondents when it came to their perception about the utility of HRIS with respect to the training function of the HR department. Whereas 50% believed that HRIS improved the training process, only 37.5% believed that the information generated form HRIS was helpful in identifying the proper time to implement a training program. Only 12.5% of the participants believed that HRIS had played a role in decreasing the cost of the training program. The conclusion which can be drawn from this feedback can indicate structural problems as there might not be a proper training needs assessment form made available. However, other important reasons can be as follows:

- The HIRS is not mature enough to have the capacity of properly incorporating the training needs of employees.
- The HRIS workers are not fully trained about the usage of HRIS as a tool to increase the efficiency of the training process.
- Since training has more strategic function as compared to administrative one, it is being ignored.
- The training needs assessment forms have not been properly developed. In order to rectify this situation, HR managers need to envision HRIS as an important component of the training process. This can be achieved by carefully assessing what the training needs of employees are and then updating those needs in the profile of each employee.

- Once the majority of the training needs have been ascertained, then a training schedule can be designed accordingly.
- This schedule would then be keyed into the employee's profile, so that HR is aware of exactly what type of training is required, the time it would take to complete it, as well as its frequency and the overall cost. This would in return also allow the HR department to monitor which employee has completed various training programs and whether that particular training helps employees in better performing their jobs. If all these appear in the employees' profiles, then not only each employee's progression, but also the streamlining of the training process which can include duration, objectives, outcomes, relevance, and effectiveness can be ascertained.

2.10 Increasing the Strategic role of HRIS

One of the major roles of HRIS system is to improve communication between HR and other departments, facilitate effective decision making make effective decisions and gain a competative advantage for the organization. Only 37.5% of the respondents surveyed believed that HRIS was fulfilling this goal. Whereas 87.5% believed that information was being shared between top administrators, only 50% of these administrators were actually using this information. Only half of the people that were surveyed thought that HRIS lived up to their expectations, whereas all 100% agreed that HRIS can and should be better utilized.

Regarding strategic consideration, it is safe to assume from the results that HRIS will continue to play a more administrative rather than a strategic role within most organizations. This trend can be reversed if the available information is disseminated to other employees within the organization. However, management has to make sure that the right information reaches the right people. In addition it costs the organization both time and money when employees have to look through stacks of information to identify which is most relevant to them. This in turn impairs the employee's ability to think strategically. This problem can be overcome by providing relevant information access to each department.

The extent to which HRIS can provide a competitive advantage to any organization is contingent on the role of the HR department within that organization. In institutions where HR is mostly confined to a personnel or employee advocate role, it is difficult to see how even the most effective HRIS can contribute towards increasing the competitive advantage of the organization.

The reports that are generated might not be user-friendly and that might be the reason why the information generated by the system is not being properly utilized to its fullest potential. The reason can be that people are not encouraged to read the reports and then make tactical decisions, based on the information provided. Overall, more needs to be done and further research needs to be conducted to discover how HRIS can be better utilized to strategically benefit the entire organization.

2.11 Conclusion

The result supports the finding that HRIS is mostly being employed as an administrative tool more than a strategic one. The holistic view of the role that HRIS can play in improving the efficiency and integration of HR department into a more strategic role was missing. The respondents could not establish a direct link between HRIS and its impact on their routine work. There was a lack of clarity as to the exact value the HIRS system would add to the organization. This relates back to the earlier literature, that the benefits of HRIS are difficult to quantify, and cannot be displayed in monetary terms. Neither cost saving, strong communication nor effective recruitment decisions were linked directly to HRIS. So even though HRIS appears to have tremendous promise it has not been fully utilized according to its potential. However, more research should be done in other sectors to see whether these finding are similar in different industries.

References

1 Armstrong, Michael (2009), A Handbook of Human Resource Management Practice (11th ed). London: Kogan Page.
2 Beer, M. and Specter, B. (1985). Corporate wide transformation in HRM, In Walton R, E. and Lawrence P. R. (Eds) HRM: Trends and Challenges. Boston, MA Harvard University Business School Press, pp. 219–253.
3 Broderick, R., Boudreau J. W. (1992). Human resource management, Information technology and the competitive advantage, Academy of Management Executive 6 (2), 1992, 7–17.
4 Ball, K. S, (2001). The use of human resource information systems: a survey, Personnel Review. 30. 677–693.
5 Beadles, Nicholas C. M. (2005). The Impact of Human Resource Information System: an Explorartory Study in Public Sector. Communications of llMA, 39–46.

6 Burbach, R. and Dundon, T. (2005). The strategic potential of human resource information systems: evidence from the republic of Ireland, Intentional Employment Relations Review, 11(1/2), 97–117.

7 Chin-Loy, C and Mujtaba, B. G. (2007). The Influence of Organizational Culture on the Success of Knowledge Management Practices with North American Companies. International Business and Economics Research Journal 6(3), 15–29.

8 Dessler, G., Griffiths, and B. Lloyd- Walker (2004). Human Resources Management, 2nd ed. Frenchs Forest, New South Wales: Pearson Education Australia, 2004, pp. 97–99.

9 Delorrae, M. and Arcand, M. (2010). HRIS implementation and deployment: a conceptual framework of the new roles, responsibilities and competences for HR professionals. International Journal of Business Information Systems, 5, 148–161.

10 DeSanctis, Gerardine (1986). Human Resource Information Systems- A Current Assessment. MIS Quarterly, 10(1), 15–27.

11 Gallagher, M. (1986). Computers in Personnel Management, Heinemann, UK.

12 Huselid, M.A., Jackson, S.E. and Schuler, R. S. (1997). Technical and Strategic Human Resource Management Effectiveness as Determinants of Firm Performance. Academy of Management Journal, 40, 171–188.

13 Hendrickson R, Anthony (2003). Human Resources Information Systems: Backbone Technology of Contemporary Human Resources. Journal of Labor Research, 24(3), 382–394.

14 Ivancevich, J. M. (2007). Human Resource Management. New York: New York: McGraw-Hill/Irwin

15 Kavanagh, M. J., Gueutal, H, G., and Tannenbaum, S, I. (1990), Human resource information systems: development and application. Boston, Mass: PWS-Kent Publications Co.

16 Kovach, K.A., Hughes, A.A., Pagan, P., & Maggitti, P. G., (2002). Administrative and strategic Advantages of HRIS. Employment Relations Today, 29, 43–8.

17 Krishnan, S., & Singh, M. (2006). Issues and concerns in the implementation and maintenance of HRIS. Issues and concerns in the implementation and. Indian institute of management Ahmedabad-380015. Research and Publication Department In its series IIMA working papers with number WP2006-07-01.

18 Lloyd, C., and Rawlinson, M. (1992). New technology and human resource management in Blyton, P, and Turnbull, P. (eds) Reassessing Human Resource Management. London Sage Publications, pp. 185–199.

19 Legge, K. (2005). Human Resource Management: Rhetorics and Realities (Anniversary Ed). Basingstoke: Palgrave MacMillan.

20 Marchiogton, M. and Wilkinson, A. (2002). People Management and Development (2nd Ed) London, CIPD.

21 Martinsons, M. G-(1994). Benchmarking human resource information systems in Canada and Hong Kong. Information & Management, 26, 305.16.

22 Mujtaba, B. G. (2011). A Cross-Cultural Comparison of Business Ethics Study with Respondents from Afghanistan, Pakistan, Iran, and the United States. International Leadership Journal. 3 (1X40–60.

23 Mujtaba, B. G. (2007). Workplace Diversity Management: Challenges, Competencies and Strategies, LIumina Press: Florida.

24 Mujtaba, Bahaudin G. and Afza, Talat (2011). Business Ethics Perceptions of Public and Private Sector Respondents in Pakistan. Far East Journal for Psychology and Business, 3(1), 01–11.

25 Mujtaba, B. G., Afza, T., and Habib, N. (2011). Leadership Tendencies of Pakistanis: Exploring Similarities and Differences based on Age and Gender. Journal of Economics and Behavioral Studies, 2(5), 199–212.

26 Noon, M. (1992) HRM: A map, model or theory? in Blyton, P. And Turnbull, P. (Eds) Reassessing Human Resource Management London: Sage Publications.

27 Ngai, E.W., and Wat, F.K. (2006). Human resource information systems: a review and empirical analysis. Human Resource Information Systems, 35, 298–314.

28 Parry, E., Tyson, S., Selbie, D., & Leighton, R. (2007). HR and Technology: Impact and Advantages, London: Charted Institute of Personnel and Development.

29 Perry, E. (2010). The benefits of using technology in human resource management, IG1 global Cranfield School of Management.

30 Schuler, R.S., Dolan, S.L. and Jackson, S. (2001). Trends and emerging issues in human resource management: global and Trans cultural perspectives - introduction. International Journal of Manpower, 22(3), 195–197.

31 Schuler, R. S., Jackson, S.E., and Storey, J. J. (2001). HRM and its link with strategic management, in: J. Storey (Ed.), Human Resource Management: A Critical Text, second ed., Thomson Learning, London. 2001.

32 Tyson, S. (1995). Human Resource Strategy: Towards a general theory of human resource management. London, Pitman.

33 Yasmeen, G., Begum, R., and Mujtaba, B. G. (June 2011). Human Development Challenges and Opportunities in Pakistan: Defying Income Inequality and Poverty. Journal of Business Studies Quarterly, 2(3), 1–12.

34 Wyatt Watson. (2002). e-HR: Getting Results Along the Journey - 2002 Survey Report Watson Wyatt Worldwide.

3

Human Resource Management in India: 'Where from' and 'Where to?'

Samir R. Chatterjee

President of the Society for Global Business and Economic Development, Professor of International Management, Curtin University of Technology, Australia

3.1 Abstract

India is being widely recognised as one of the most exciting emerging economies in the world. Besides becoming a global hub of outsourcing, Indian firms are spreading their wings globally through mergers and acquisitions. During the first four months of 1997, Indian companies have bought 34 foreign companies for about U.S. $11 billion dollars. This impressive development has been due to a growth in inputs (capital and labour) as well as factor productivity. By the year 2020, India is expected to add about 250 million to its labour pool at the rate of about 18 million a year, which is more than the entire labour force of Germany. This so called 'demographic dividend' has drawn a new interest in the Human Resource concepts and practices in India. This paper traces notable evidence of economic organisations and managerial ideas from ancient Indian sources with enduring traditions and considers them in the context of contemporary challenges.

3.2 Introduction

Over many centuries India has absorbed managerial ideas and practices from around the world. Early records of trade, from 4500 B.C. to 300 B.C., not only indicate international economic and political links, but also the ideas of social and public administration. The world's first management book, titled 'Arlhashastra', written three millennium before Christ, codified many aspects of human resource practices in Ancient India. This treatise presented notions of the financial administration of the state, guiding principles for trade and commerce, as well as the management of people. These ideas were to be embedded in organizational thinking for centuries (Rangarajan 1992, Sihag 2004). Increasing trade, that included engagement with the Romans, led to widespread and systematic governance methods by 250 A.D. During the next 300 years, the first Indian empire, the Gupta Dynasty, encouraged the establishment of rules and regulations for managerial systems, and later from about 1000 A.D. Islam influenced many areas of trade and commerce. A further powerful effect on the managerial history of India was to be provided by the British system of corporate organization for 200 years. Clearly, the socio cultural roots of Indian heritage are diverse and have been drawn from multiple sources including ideas brought from other parts of the old world. Interestingly, these ideas were essentially secular even when they originated from religious bases.

In the contemporary context, the Indian management mindscape continues to be influenced by the residual traces of ancient wisdom, as it faces the complexities of global realities. One stream of holistic wisdom, identified as the Vedantic philosophy, pervades managerial behavior at all levels of work organizations. This philosophical tradition has its roots in sacred texts from 2000 B.C. and it holds that human nature has a capacity for self transformation and attaining spiritual high ground while facing realities of day to day challenges (Lannoy 1971). Such cultural based tradition and heritage can have a substantial impact on current managerial mindsets in terms of family bonding and mutuality of obligations. The caste system, which was recorded in the writings of the Greek Ambassador Megasthenes in the third century B.C., is another significant feature of Indian social heritage that for centuries had impacted organizational architecture and managerial practices, and has now become the focus of critical attention in the social, political and legal agenda of the nation.

One of the most significant areas of values and cultural practices has been the caste system. Traditionally, the caste system maintained social or organisational balance. Brahmins (priests and teachers) were at the apex, Kshatriya (rulers and warriors), Vaishya (merchants and managers) and Shwdra (artisans and workers) occupied the lower levels. Those outside the caste hierarchy were called 'untouchables'. Even decades ago, a typical public enterprise department could be dominated by people belonging to a particular caste. Feelings associated with caste affairs influenced managers in areas like recruitment, promotion and work allocation (Venkatranain & Chandra 1996). Indian institutions codified a list of lower castes and tribal communities called 'scheduled castes and scheduled tribes'. A strict quota system called, 'reservation' in achieving affirmative equity of castes, has been the eye of political storm in India in recent years. The central government has decreed 15 per cent of recruitment is to be reserved for scheduled castes, and a further seven and half per cent for scheduled tribes. In addition, a further 27 per cent has been decreed for other backward castes. However, the liberalisation of markets and global linkages have created transformation of attitudes towards human resource (HR) policies and practices (Khalilzadeh-Shirazi & Zagha 1994, Gopalan & Rivera 1997). Faced with the challenge of responding to the rationale of Western ideas of organisation in the changing social and economic scenario of Indian organization, practitioners are increasingly taking a broader and reflective perspective of human resource management (HRM) in India.

This manuscript has three main parts. In the first part is provided an overview of important historical events and activity that has influenced contemporary managerial tenets, the second part of the manuscript describes the emerging contemporary Indian HRM practices and indicates some interesting challenges. Much of the second part is also summarised on four informative Figures, The concluding section, the third part of the manuscript, succinctly integrates the two preceding parts.

3.3 Value of Context of HRM in India

The managerial ideologies in Indian dates back at least four centuries. Arthashastra written by the celebrated Indian scholar-practitioner Chanakya had three key areas of exploration, 1) public policy, 2) administration and utilisation of people, and 3) taxation and accounting principles (Chatterjee 2006). Parallel to such pragmatic formulations, a deep rooted value system, drawn from the early Aryan thinking, called vedanta, deeply influenced the societal and institutional values in India. Overall, Indian collective culture had an interesting individualistic core while the civilisational values of duty to family, group and society was always very important while vedantic ideas nurtured an inner private sphere of individualism.

There has been considerable interest in the notion that managerial values are a function of the behaviours of managers. England, Dhingra and Agarwal (1974) were early scholars who contended that managerial values were critical forces that shape organisational architecture. The relevance of managerial values in shaping modern organisational life is reflected in scholarly literature linking them to corporate culture (Deal and Kennedy 1982), organisational commitment and job satisfaction (O'Reilly, Chatham and Caldwell 1991), as well as institutional governance (Mowday, Porter and Steers 1982). Thus, understanding the source of these values and in particular societal work values (which link the macro-micro relationships and in turn organisational practices) had become a popular line of enquiry, and a great deal of evidence has been presented to support the importance of national culture in shaping managerial values. One of the most widely read formulations of this literature is the seminal work of Hofstede (1980) who popularised the notion of clustering culture in generic dimensions such as power distribution, structuring, social orientation, and time horizons. In turn, these dimensions could be employed to explain relevant work attitudes, job incumbent behaviours and the working arrangements within organisational structures. Two of these dimensions were individualism and collectivism.

The traditional social ethos from the ancient roots, which was developed over centuries, underwent profound transformation during the British rule. Consequently, in the contemporary context multiple layers of values (core traditional values, individual managerial values, and situational values) have emerged (Chatterjee & Pearson 2000). Though the societal values largely remain very much anchored in the ancient traditions they are increasingly reflecting corporate priorities and values of global linkages. But in the arena of globalisation where priorities of consumerism, technological education, mass media, foreign investment and trade union culture predominate, newer tensions are becoming evident. For instance, contemporary Indian multi national companies and global firms in India have started shifting their emphasis to human resources with their knowledge and experience as the central area of attention in extending new performance boundaries (Khandekar & Sharma 2005). Considerable research evidence attests to this trend with particular relevance to greenfield organisations with little or no historical baggages in their organisational culture (Settt 2004, Roy 2006).

Within Indian traditions the choice of Individualistic or collectivistic behavior depends on a number of culturally defined variables. The dynamics of these variables are underpinned through three key elements guiding Indian managerial mindscapes. These three constructs are Desh (the location), Kaal (the timing), and Patra (the specific personalities involved). Sinha and Kunungo (1997) claim that the interaction of these three variables determines the guidelines for decisions cues. This managing or nurturing of the outer layer of collectivism in an inner private sphere of individualism is expressed in

DECISIONAL	Desh	Kaal	Patra
CUES	(Place)	(Timing)	(Actors)
SPIRITUAL	Sattava guna	Tamas guna	Rajas guna
ORIENTATION	(Virtue focus)	(Negative focus)	(Action focus)
INTERPERSONAL	Sradha	Sneha	Bandhan
RELATONS	(Upward respect / Loyalty	(Downward affection)	(Bonding)

Figure 1   Behavioral Anchors in Indian Organisational Life

Figure 1 which demonstrates the behavioural anchors in Indian organisational life.

Figure 1 also presents another powerful insight of the Indian tradition of the notion of 'Guna' dynamics. According to Sharma (1996), this culture based framework, which has three types of gunas (attraction), is being increasingly used in employee assessment and organisational team building strategies. The contention is that each guna is a separate contribution to the core of human personalities. The Sattava (or truth orientation) is the sentiment of exalted values in people, organisations or society. Alternatively, the Tamasik guna depicts a negative orientation which can be expressed behaviourally as ignorance, greed or corruption. Those individuals with a Rajasik guna are inherently driven by a desire to make a worthwhile contribution to their surroundings. Collectively, these spiritual orientations, which manifest as Sattava, Tamas or Rajas gunas, articulate as positive or negative HRM functions such as leadership, motivation or other institutional behavioural activity. The third row of Figure 1 highlights the linking of HRM trends to socio cultural roots. The culture of Sradha (upward loyalty) and Sneha (mentoring with affection) outline the behavioural anchors derived from the civilisational roots. The acceptance of 'Sradha' by youngers and the display of 'Sneha' by the seniors have been the root of sustainability of all types of Indian oragnisations. This has a striking similarity to the concepts of 'oyabun' and 'kobun' in the Japanese cultural context.

3.4 Contemporary India

In a recent survey of Indian CEO's, it was suggested that Indian managerial leaders were less dependent on their personal charisma, but they emphasised logical and step by step implementation processes. Indian leaders focused on empowerment and accountability in cases of critical turnaround challenges, innovative challenges, innovative technology, product planning and marketing or when other similar challenges were encountered (Spencer, Rajah, Narayan, Mohan & Latin 2007), These social scientists contend.

Leaders in other countries often tell about why they chose a peculiar person for a certain role per task, detailing the personal characteristics that made that person right for that situation. They may also consider, in detail, how an assignment would help someone grow and develop their abilities. In general, Indian leaders simply did not discuss how they matched particular people to certain roles or tasks, nor did they usually consider in detail how the personal characteristics of individuals might shape or inform the best way to influence that person. (Spencer, et al 2007:90).

3.5 Indian HRM in Transition

One of the noteworthy features of the Indian workplace is demographic uniqueness. It is estimated that both China and India will have a population of 1.45 billion people by 2030, however, India will have a larger workforce than China. Indeed, it is likely India will have 986 million people of working age in 2030, which well probably be about 300 million more than in 2007. And by 2050, it is expected India will have 230 million more workers than China and about 500 million more than the United States of America (U.S.). It may be noted that half of India's current population of 1.1 billion people are under of 25 years of age (Chatterjee 2006). While this fact is a demographic dividend for the economy, it is also a danger sign for the country's ability to create new jobs at an unprecedented rate. As has been pointed out by Meredith (2007).

When India's young demographic bubble begins to reach working age, India will need far more jobs than currently exist to keep living standards from declining. India to-day doesn't have enough good jobs for its existing workers, much less for millions of new dues. If it cannot better educate its children and create jobs for then once they reach working age, India faces a population time bomb: The nation will grow poorer, and not richer, with hundred of millions of people stuck in poverty. (p.133).

With the retirement age being 55 to 58 years of age in most public sector organisations, Indian workplaces are dominated by youth. Increasing the retirement age in critical areas like universities, schools, hospitals, research Institutions and public services is a topic of considerable current debate and agenda of political parties. The divergent view, that each society has an unique set of national nuances, which guide particular managerial beliefs and actions, is being challenged in Indian society. An emerging dominant perspective is the influence of globalization on technological advancements,

business management, education and communication infrastructures is leading to a converging effect on managerial mindsets and business behaviours.
And when India embraced liberalisation and economic reform in the early
1990s, dramatic changes were set in motion in terms of corporate mindsets and
HPM practices as a result of global imperatives and accompanying changes in
societal priorities. Indeed, the onset of a burgeoning competitive service sector
compelled a demographic shift in worker educational status and heightened
the demand for job relevant skills as well as regional diversity. Expectedly,
there has been a marked shift towards valuing human resources (HR) in Indian
organisations as they become increasingly strategy driven as opposed to the
culture of the status quo. Accordingly, competitive advantage in industries like
software services, Pharmaceuticals, and biotechnology (where India is seeking
to assert global dominance), the significance of HRs is being emphasised.
These relativities were demonstrated in a recent study of three global Indian
companies with (235 managers) when evidence was presented that positively
linked the HRM practices with organisational performance (Khandekar &
Sharma 2005). In spite of this trend of convergence, a deep sense of locality
exists creating more robust 'cross vergence' in the conceptual as well as
practical domain. Figure 2.

Figure 2 presents the key drivers for contemporary Indian HRM trends. In
Figure 2 there are four external spheres of intervention for HRM professionals
and these spheres are integrated in a complex array within organisational
settings. The intellectual sphere, which emphasises the mindset transaction in
work organisations, has been significantly impacted by the forces of globalisation, Indeed, Chatterjee and Pearson (2000) argued, with supporting empirical
evidence from 421 senior level Indian managers, that many of the traditional
Indian values (respect for seniority, status and group affiliation) have been
complemented by newer areas of attention that are more usually linked to
globalisation, such as work quality, customer service and innovation. The most
important work related attribute of the study was the opportunity to learn new
things at work. Such cross verging trends need to be understood more widely
as practitioners face a new reality of human resource development of post
industrial economic organizations.

The other three spheres, of Figure 2, namely the emotional, the socio cultural and the managerial domains are undergoing, similar profound changes.
For instance, the socio cultural sphere confronts the dialects of the national
macro level reform agenda as well as the challenge of innovating by addressing
the hygiene and motivational features of the work place. Consequently, this
sphere, which is underpinned by the anchors of Sradha and Sneha, has the

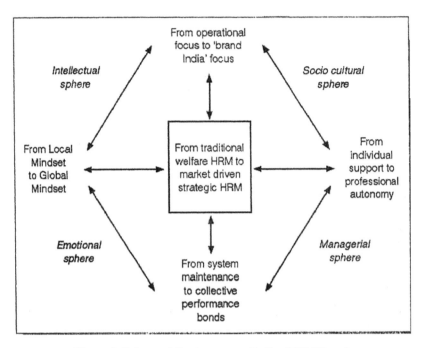

Figure 2 Drivers of Contemporary Indian HRM Trends

opportunity to leverage work setting creativity in dimensions of autonomy, empowerment, multiskilling and various types of job design. And the emotional sphere, which focuses on creativity and innovation to encapsulate the notions of workplace commitment and collaboration as well as favourable teamwork, brings desirable behavioural elements of transparency and integrity into organisational procedures and practices. The managerial sphere provides the mechanisms for shifting mindsets, for in Indian organisations HRM is viewed to be closely aligned with managerial technical competency. Thus, understanding of the relativity of HRM to strategic intended organisational performance is less well articulated in Indian firms. The current emphasis of reconfiguring cadres (voluntary and nonvoluntary redundancy schemes), downsizing, delayering and similar arrangements will become less relevant as holistic perspectives gain ground. A hallmark of future Indian workplaces is likely to be a dominant emphasis on managerial training, structural redesign and reframing of institutional architectures to achieve enterprise excellence. Thus, a primary role of Indian managers will be to forge new employment and industrial relationships through purposeful HRM policies and practices.

HRM Practices	Observable Features
Job Description	Percentage of employees with formally defined work roles is very high in the public sector.
Recruitment	Strong dependence on formal labour market. Direct recruitment from institutions of higher learning is very common amongst management; engineering and similar professional cadres. Amongst other vehicles, placement agencies, internet and print media are the most popular medium for recruitment.
Compensation	Strong emphasis on security and lifetime employment in public sector including a range of facilities like, healthcare, housing and schooling for children.
Training and Development	Poorly institutionalised in Indian organisations. Popularity of training programmes and their effect in skill and value development undeveloped.
Performance Appraisal	A very low coverage of employees under formal performance appraisal and rewards or organisational goals.
Promotion and Reward	Moderately variable across industries. Seniority systems still dominate the public sector enterprises. Use of merit and performance limited mostly to globally orientated industries.
Career Planning	Limited in scope. The seniority based escalator system in the public sector provides stability and progression in career. Widespread use of voluntary retirement scheme in public sector by high performing staff. Cross functional career paths uncommon.
Gender Equity	Driven by proactive court rulings, ILO guidelines and legislature provisions. Lack of strategic and inclusion vision spread.
Reservation System	The central government has fixed 15 per cent reservations for scheduled castes, 7.5 per cent for scheduled tribes and 27 per cent for backward communities. States vary in their reservation system.

Figure 3 Key HRM Practices in Indian Organisations

In Figure 3 is presented a variety of HRM practices that are being employed in Indian organisations.

3.6 IR Challenge

The Indian IR system has two main features. First, is the absence of the provision to recognise a union as a representative or agent for collective bargaining. Second, is the total dominance of government in regulating the industrial relations (IR) domain. Though it is relatively easy for members of a work organisation to be registered as a union under the law, it does not lead to the legal recognition by the employer in dispute resolution or bargaining process. This contention was made by Kuruvilla (1996) over a decade ago.

In terms of collective bargaining, industry wide bargaining occurs in certain industries where the employers are organized, but bargaining otherwise is decentralised to enterprise level. Although there are no restrictions on the subjects of bargaining, the Industrial Disputes Act of 1947 restricts the ability of employers to lay off or retrench employees or to close business. (p.635).

Indian industrial relations have evolved from political roots and labour market demands. An unique feature of Indian IR has been the dominance of political parties sponsoring unions. Union membership has been the most popular breeding ground for politicians, and political leaders have enjoyed the use of union platforms. Such politicisation has generated conflicts and rivalry creating mayhem and the hurting of labour interest. Nevertheless, in spite of wage determination by central government boards, and ad hoc industrial awards, enterprise level bargaining has yielded. positive outcomes. Interestingly, during the 1970s in a period of the highest number of strikes, the registered number of unions grew fivefold. But a decade later profound economic and political reform movement saw a new direction in the trade union movement. A section of scholarly trade union leaders began to incorporate new global thinking in the union outlook. Since the 1980s, the Indian industrial relations culture has been considerably impacted by the intensification of globalised markets. During this time and beyond, there has been a clear departure from traditional personnel management. The shift has not only been in the general tone, but in the substantive visions. Adjustment to the global imperatives of an emerging service sector, sunrise industries, and demographic shifts in competencies has given rise to new thinking. In spite of most of the Indian labour laws being entrenched in a world view that is very different to the current realities, and the obvious urgency for them to be updated to incorporate more flexible, competitive work systems, the built in rigidities are still proving a formidable obstacle. The most alarming issue in the HR and IR context is the lack of job opportunities outside urban areas where more than 70 per cent of the population lives. As has been pointed out by Meredith (2007).

While Indian university graduates line up for jobs that can propel them into newly vibrant middle class, per India's rural and urban poor, change has been interminably delayed. Expectations, like incomes, are rising across India, and not just for those working in call centres. Even as the New India cohort thrives, much of the rest of India is making much slower gains or even being left behind, creating social and political tensions that cloud India's impressive strides forward. The lowest paid workers in the off shoring industry those working in the call centres earn median wages of $275 a month. But most Indians still earn less than $60 a month or just $2 a day. (p.125).

3.7 Technical Services Recruitment and Retention

There has been a dramatic shift in the expectations of employees in the organised and globally linked sectors of the economy. An unprecedented rise in the disposable income coupled with a declining dependency ratio, has led to young professionals becoming extremely mobile. The problem is critically evident in the off shoring industry where the average retention period of an employee is considered to be around six to eight months. And the retention of senior level executives is an additional challenge. The attrition rates are highest in information technology (IT) (30–35%), business process outsourcing (BPO) (35–40%), insurance (35–40%), retail and fast moving consumer goods (FMCG) (20–30%), and manufacturing and engineering (10–15%) (Chatterjee 2006).

Over the past decade, there has been a sea change in the area of Indian technical services and the associated HRM practices of recruitment and retention. While the higher education system in the country has remained overwhelming poor in infrastructure and weak in becoming revitalised to grapple with the global imperatives, there has been a mushrooming of private educational institutions. The recruitment problem is further deepened by the emergence of a new culture of 'job hopping' amongst employers who can demonstrate their world class competencies. This phenomenon of turnover has seen a chain reaction in entry level salaries, and an increase in graduates has created significant social and economic disruption to the Indian labour market. A likely scenario from this rampant activity is that the Indian HR scene will be negatively impacted in the next decade unless the deregulation and autonomy of the higher education sectors is initiated somewhat immediately. An example of this widening gap between the university system and market need has become a serious impediment in several new industries in India. For an example, it has been reported in the popular press (Time 2007), "...out of 13 million people who applied to work at IT company Infosys last year, just 2% were qualified indicating a sign of stress in the university system that graduates 2.5 million a year".(p.33).

One of the most concerning issues for HR managers in India is the high staff turnover. In industries like call centres, staff attrition is the single biggest issue. The industry has grown from zero employment to an employer of quarter of a million young English speaking, well educated and ambitious people. The point is well made by Slater (2007), who wrote.

Attrition is highest in traditional customer service jobs, where young people find themselves having to spend all night on the phone, often with irate callers. In other areas, such as claims processing or accounting, the turnover rate is much lower. More worrying for many companies is the 'merry go round' in supervising and management jobs, as new centers are only too willing to pay higher salaries to hijack experienced staff- (p.34).

The issue of retention is much more critical in the high value adding BPO sector Such as R&D activities. This $40 billion industry has one of the highest attrition rates of around 20 to 25 per cent. The service laden BPO and Hord industry have the highest attrition rates. Many companies are developing innovative incentive packages in countering this job hopping phenomenon. Figure 4 illustrates some of these initiative by leading companies in India.

A dramatic shift in recruitment practices has been taking place as globally pretend Indian companies as well as global technical services rivals have made India a battlefield of recruitment for the best workers. For example, IBM's workforce in India has more than doubled in two years to a cadre of 53,000. This outcome has come with the elimination of 20,000 jobs in high cost markets like the U.S., Europe and Japan, The R&D centre of IBM is staffed by 3,000 world class engineers and is being recognised for its ability to innovate on all areas from simple processes, softwares, semiconductors as well as supercomputers. It is interesting to note that IBM has dominated the recruitment market in technical services in India during 2006. This leading company recruited 10,000 employees out of a total of 25,000 people who were recruited to the technical services industry. The prominence of IBM as an employer of technically qualified personnel has been acknowledged in the popular press' (Business Week. 2007).

In Pune, a rapidly developing IT centre near Mumbai, the company has been dispatching vans with signs saying, 'IBM is hiring', to the gates of the rivals at lunch time. Their hit rate is pretty good laments a manager at a tech firm that has lost employees to IBM.

Name of the Company	Retention Strategy	Impact
Tata Consulting Services (TCS)	• A choice of working in over 170 offices across 40 countries in a variety of areas. • Paternity leave for adoption of a girl child • Discounts on group parties	• Significant impact on job hopping achieved
ICICI Bank	• Identification of potential talented staff • Alternative stock options • Quicker promotion	• Have been able to achieve higher retention rate
WIPRO	• 'Wings Within' programme where existing employees get a chance to quit their current job role and join a different firm within WIPRO	• Has led to a higher retention rate
INFOSYS	• Fostering a sense of belongingness, creative artistic and social activities for the employees and their families. • Initiating one of the best 'corporate universities' in the world	• Moderate Retentions rate increase achieved
Microsoft-India	• Excellent sporting and wellness facilities • Employees allowed to choose flexible working schedule • Moving people across functions and sections in assisting employees find their area of interest	• Struggling to minimise job hopping
Mahindra & Co	• Culture changes valuing innovation and talent over age and experience • Instituionalising a practice called 'reverse mentoring' where young people are given opportunities of mentoring their seniors	• Stabilised job hopping significantly

Figure 4 Examples of Retention Strategies for young Professionals in India's BPO and Services Sectors

3.8 Conclusion

The World Competitiveness Report rated India's human resource capabilities as being comparatively weaker most Asian nations. The recognition of world class human resource capability as being pivotal to global success has changed Indian HRM cultures in recent years. While the historical and traditional roots remain deeply embedded in the subjective world of managers, emphasis on objective global concepts and practices are becoming more common. Three very different perspectives in HRM are evident. Firstly, Indian firms with a global outlook; secondly, global firms seeking to adapt to the Indian context; and thirdly, the HRM practice in public sectors undertakings (PSVS). As the Indian economy becomes more globally linked, all three perspectives will move increasingly towards a cross verging strengthening. Interestingly, within the national context, India itself is not a homogenous entity. Regional variations in terms of industry size, provincial business culture, and political issues play very relevant roles. The nature of hierarchy, status, authority, responsibility and similar other concepts vary widely across the nations synerging system maintenance. Indeed, organisational performance and personal success are critical in the new era.

References

1 Business Week. (2007). A Red-HA Big Blue in India, Sept 3. Avaliable: http://businessweek.com/magazine/content/07_36/b4048052.htm

2 Chatteijee, S.R. (2006). Human resource management in India, In A. Nankervis, Chatterjee, S.R. & J. Coffey (Eds.), Perspectives of human resource management in the Asia Pacific (41–62). Pearson Prentice Hall: Malaysia.

3 Chatterjee, S.R., & Pearson C.A.L. (2000). Indian managers in transition: Orientations, work goals, values and ethics. Management International Review, 40(1), 81–95.

4 Deal, E., & Kennedy, AA (1982), Corporate culture. Reading, MA,: Addison-Wesley.

5 England, G.W., Dhingra, O.P., & Agarwal, C.N. (1974). The manager and the man: A crosscultural study of personal values. Kent, Ohio: The Kent State University Press.

6 Gopalan, S., & Rivera, J.B. (1997). Gaining a perspective on Indian value orientations: Implications for expatriate managers. International Journal of Organizational Analysis, 5(2), 156–179.

7 Lannoy, R. (1971). The speaking Tree: A study of the Indian society and culture. Oxford: Oxford University Press.

8 Khalilzadeh-Shirazij J., & Zagha, It (1994). Achievements and the agenda ahead. The Columbia Journal of World Business, 29(1), 24–31.

9 Khandekar, A., & Shanna, A. (2005). Managing human resource capabilities for sustainable competitive advantage: An empirical analysis from Indian global organization. Education & Training, 47(47/48), 628–639.

10 Kuravilla, S. (1996). Linkages between industrial nation strategies and industrial relations/human resource policies: Singapore, Malaysia, the Philippines and India, Industrial & Labour Relations Review, 49(4)1 635–658.

11 Meredith, R. (2007). The elephant and the dragon: The rise of India and China and what it means for all of us. New York: W,W JMorton & Co.

12 Mowday, R.T., Porter, L.W., & Steers, RJML (1982). Employee-organization linkages: The psychology of commitment, absenteeism and turnover. New York: Academic Press.

13 O'Reilly, CA., Chatham, JA., & Caldwell, D.F. (1991). People and organizational culture: A profile comparison approach to assessing person-organization fit. Academy of Management Journal, 34(3), 487–516.

14 Rangarajan, LN. (1992). He Arthashastra. New Dheli: Penguin Books

15 Roy, R. (2006). Quality of work life as a determinant of mental health: SCMS Journal of Indian Management, 3(2)

16 Sett, P.K. (2004). Human resource management and firm level restructuring: The south Asian drama, Research and Practice in Human Resource Management 12(1), 1–18.

17 Singh, K. (2003). Strategic HR, Orientation and Firm Performance in India. International Journal of Human Resource Management, 14(4), 530–543.

18 Sihag, B. (2004), Kautilya on the scope and methodology of Accounting, organizational design and role of ethics in ancient India. The Accounting Historians Journal, 31(2), 125–148.

19 Sinha, J., & Kannungo, R. (1997), Context sensitivity and balancing in organizational behaviour. International Journal, of Psychology, 32(1), 93–105.

20 Slater, J. (2004). Job-hopping central: Far East Economic Review, 8(1X34.

21 Spencer, S., Rajah, T., Narayan, S., Mohan, S., & Latin, G. (2007). The Indian CEO: A portrait of excellence. New Delhi: Response Books.

22 Time. (2007). Special Report: 60 years of Independence, 170(6), 4–42.
23 Venkatratnam, C., & Chandra, V. (1996). Sources of diversity and the challenge before human resource management in India. International Journal of Manpower, 17(4/5), 76–96.

4

The Role of Information Technology in Human Resources Management

Ing. Iveta Gabcanova

Univerzita Tomase Bative Zline, Fakulta managementu a ekonomiky, Ceska republika

4.1 Abstract

In nowadays top leaders fully realize the power of information technology (IT) tools for reaching business targets. The utilization of IT tools help not only to fulfill defined company's goals but to optimize the work processes as well. Trends and results of the contemporary studies constantly confirm contribution of the IT tools in Human Resources (HR) area i.e. to accomplish assigned HR tasks by using the source of IT capabilities. The following paper gives a brief overview about possibilities of IT usage in HR field for measuring and tracking human capital and using the HR information system generally.

Keywords: Information Technology. Human Resources Management, Company's goals, Recruiting, Idea management, Human Resources development.

4.2 Introduction

There is no underestimation of importance and effect of the Human Resources management at all. Lately, management of Human Resources and its needs are becoming the center of the attention of each individual employer in every organization. The orientation of company on human resources starts to be one of the key tasks of a strategic management and Human Resources play an important role in all strategic decisions. Managers of Human Resources ask for more strategic position of their department within the organization with the aim to get to the essence of the problem how to manage, to motivate and to increase

65

the performance of organization. The importance of human potential for company increases proportionally with the speed of changes which appear in the business area because human capital represents a basic qualitative parameter of fruitfulness of any changes. Following that, Human Resources Management (HRM) must aim at achieving the competitiveness of the company in the field of HR by means of providing constant educational and training programs for personal development of employees. It has been Scientology proven that one of the supporting pillars which can contribute to the fulfillment of the personal policy is the usage of IT technologies in HR. Information and Communication Technologies (ICT) - a catchall term for techniques associated with mobile communication, internet, new media and PCs - allow companies to improve their internal processes, core competencies, organizational structures as well as relevant markets on a global scale. ICT is spreading throughout every sector of the economy and has implications for almost every enterprise (Helfen and Kruger, 2002).

Human resource processes should be focused on the strategic objectives. These strategies are led to prepare an IT strategic plan that in turn translates into an appropriate human resource strategic plan in the field of IT as the Figure 1 depicts (Sameni and Khoshalhan, 2006).

Source: SAMENI, M.K., KHOSHALHAN, F. Analysis of Human Resource Development for Information Technology and E-Commerce in Iran. IT Department, Faculty of Industrial Engineering, K.N.Toosi University of Technology, Tehran, Iran. 2006, p. 1 190.

IT plays a critical role in leveraging and complementing human and business resources (Powell and Dent-Micallef, 1997). The importance of using the HR - IT tools, the authors express as follows: "in organizations, despite increasing needs for technological advancement, human and cultural factors play a more important role than before. However, technology is often seen

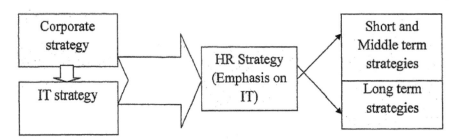

Figure 1 The Framework of IT Human Resources Plan

by management as essential means to compete in the global market. To that technology, including information technology, brings the desired results, the most important issue for an organization is how to manage the technology, with respect to human organizational aspects, how to analyze and understand human factors guided by the norms, shared beliefs, and assumptions of the organization, as well as by individual' unique values-all together known as "culture." (Zakaria and Yusof, 2001)

Information systems in HR can (Armstrong, 2002):

- provide better services to line managers,
- serve as a pipeline connecting a personal policy and personal processes in all organization and thus facilitate personal management in the company,
- provide important data for a strategic personal decision-making and enable a quick acquiring and analysis of information for HR assistants,
- reduce cost labors at performance of personal activities.

The study called "Effects of HRM practices on IT usage" (Lee, 2009) shows that organizations use technologies for HR field such as employee participation, clearly defined jobs and extensive formal training. On the other hand, according to survey, companies which are using external IT capability, only internal career opportunities used IT tools, HR function used to employ IT for administrative processes, primarily payroll processing, with little attention being paid to so-called transformational HR practices (DeSanctis, 1986), Nevertheless the results present in paper (Bondarouk and Ruel, 2009) that "in 2006, as the CedarCrestone 2006 HCM Survey shows, companies broadened the scope of HRM applications: although administrative e-HRM was still the most popular application (62% of surveyed companies), companies reported an increasing use of strategic applications like. talent acquisition services (61%), performance management (52%), or compensation management (49%)."

4.3 Direction of the Research

The main aims of an empirical study performed within multinational manufacturing companies with over five-hundred employees is to identify the level of using the HR IT tools in companies, factors on which management should make stress during the implementation of the IT tools and simultaneously find out which IT tools can be used in scope of HRM. The research is compiled by forty questionnaire surveys and consecutively by direct observation in the firms and structured interviews with fifteen HR managers.

The anonymous questionnaire and interviews were focused on the following fields:

- IT tools usage in HR field,
- Efficiency of HR IT tools usage,
- Management's support of the HR information systems application,
- Advantages and disadvantages of HR IT tools usage,
- Implementation of HR IT tools,
- HR IT tools versus HR strategic goals.

4.4 Research Results

The questionnaire reveals that 98 % of the respondents use IT tool in HR field. Concerning the efficiency of HR IT tools usage 66,7 % of respondents confirmed that HR IT tools help in everyday work considerably and 33,3 % of the respondents point out that HR IT tools support daily work moderately.

Further the survey, shows that 22 % of respondents confirmed supporting the reaching of HR goals significantly by using of HR information system and 50 % of respondents answered that HR IT tool support HR goals moderately and 25 % slightly and 3 % not at all. The results are depicted in the graph:

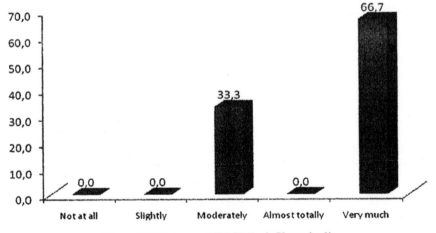

Figure 2 Efficiency of HT IT Tools Usage in %

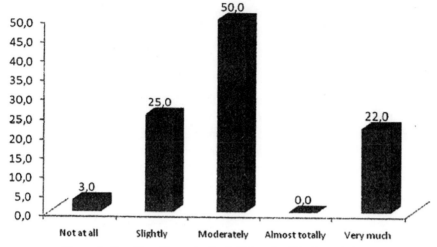

Figure 3 Reaching Strategic Goals Using the HR IT Tools in %

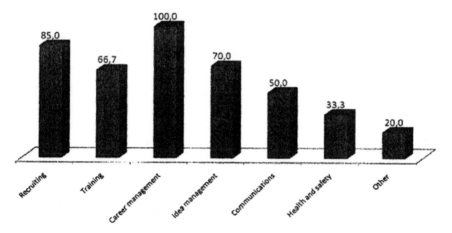

Figure 4 IT Tools Usage in HR Field in %

To evaluate the using of IT tools in HR management the respondents answered on the following question "In which HR area the IT tools are used in your company? The results are shown in the graph:

Based upon the interviews with HR managers, firstly management has to clearly define which core HR processes are to be integrated and transferred into HR IT systems. This is one of the crucial factors for effective implementation and accomplishment of HR IT usage. Consequently, it must be defined which IT tools fulfill described HR demands. Secondly, HR managers state that main advantages of using HR IT tools are time saving, work efficiency, keeping employees fully informed. The research reveals that HR managers should make stress on HR goals, the HR budget in the beginning of the Implementation of the HR IT tools. The IT tools should be user friendly, service efficient and reduce HR transactions that add no value. HR managers confirmed that usage of HR information system is essential and it makes their work easier on the one hand but it requires effective HR IT tools on the other hand. Further the interviews were focused on the three HR fields where companies use the HR information systems according to online questionnaire (Fig 3). The usage, benefit and add value of HR IT tools in firms are shortly described.

4.5 The HR Role in IT Human Resources Development

Career planning tool is a generic, learning, knowledge-based system that helps top leaders to manage the personal development and path career of employees. One of the most important online supports within Human Resources is tracking the Human Resources Development Core Processes. The tool should provide all necessary information about individual succession planning of employee, next development measures or evaluation of overall performance and review of the potential. Experience from many companies points out that the essential part of the tool is agreement of individual goals between employee and supervisor which should be directly connected to a business target. This is also connected with the fact that companies use determination of individual goals for employees in the full range in order to reach strategic goals of the company. One big advantage of the tool is that entered data are stored in database which provides attainability and visibility of the history anytime and simultaneously online system can provide a considerable cost saving. The tool should also include the reporting, it means provide information about needed trainings for employees, report about ideas for further employee's personal development. The above mentioned reports make easier the work of the Human Resources department.

4.6 The HR Role in IT Recruiting

In nowadays, managers realize that human capital has become the last competitive benefit and IT recruiting can broadly support efficient hiring together with forming the workforce. In order to attract the best candidates, it is vital that both HR and IT departments cooperate together. The HR role in IT recruiting is of key importance, including time of crisis. The war of talents carries on, despite the current crisis. The HR IT tools can support hiring and retaining a high potential. It begins with launching the career website what is a very good promotional tool. Employer can present all necessary Information related to job, careers or personal development of each applicants there. The career website should focus not only on potential employees, but target group should also include the students, graduates or school pupils in order to have a possibility to "raise" new employees for the future. The cooperation with students can save the costs especially if company needs specialists for future business needs. One big advantage which companies should use via career website is to have opportunity to apply for a job online. Additionally, company can present information about corporate culture or environment. Further; it can be connected to the internal recruiting track system which can help to HR generalist in communication with candidates to organize the selection processes. System should cover the whole application process, from position approval to candidate approach.

4.7 The HR Role in IT Idea Management Tool

To keep high performance culture, company should maintain a continuous improvement of core processes and activities. The tool which supports Idea Management processes should provide how many improvement proposals are submitted by the employees. It can also provide the information of saved revenues and costs by implementation of improvement proposals. The main purpose of online support in Idea management is to build a tool which gives the opportunity to all employees to present their applicable and well-founded ideas on the one hand and on the other hand company becomes more competitive. The above mentioned IT support carries the advantages as follows:

- Save money,
- Avoid costs or
- Improve process performance.

Considerable benefit is an immediate possibility to create various reports and finally to see if company fulfils the target of a key performance indicator. One of the last advantages of IT online system is that it helps to provide availability of improvement proposals anytime and thus enables to avoid paper form. Above mentioned gives overview how technologies can assistance HR processes in order to reach the business targets.

4.8 Discussion

The survey confirms that companies use HR IT tools and should contain all HR processes which will sustain all parts of HR it means from "Recruit to Retire" functions within the company. The research presents that the importance of HT - IT usage is getting more and more important not only due to the fact that HR productivity increases but at the same time, the value of the organization increases, including the most important asset -Human capital. First of all implementation HR IT tools requires to perform analysis which provides information about benefit of IT usage. Also during the implementation, attention must be paid to the protection of sensitive data about employees. If the tools are implemented on higher level, it can bring cost saving for the company, reduce process time, quality of work and finally the major benefit is the contribution to the strategic development of the company. It is essential that the tools are user friendly for all employees, provides data and reports necessary for the employees' jobs. Next point which has to be taken into consideration is that implementation of HR - IT tools can increase the costs e.g. purchase of the facilities, i.e. companies have to be aware of it during preparation of budget. Nevertheless, HR IT tools must help all employees because incorrect implementation may bring negative consequences e.g. the employees may understand the usage of tools as an inevitable evil which does bring any effect neither them nor the company.

4.9 Conclusions

Globalization brings the requirement to think how IT can contribute to fulfillment of business strategy in the frame of Human Resources management in order to steer the business towards excellence and reach competitiveness in the market. Despite the fact questionnaire revealed that information technology sustains reaching the HR goals moderately, precise plan of implementation of HR information system can significantly support HR strategy in the company to attain defined key performance indicators (KPI). The project should mainly contain what has to be achieved it means how IT tools can support the KPI and which kind of HR processes IT tools should sustain. The research will be extended by further survey.

References

Books

1 ARMSTRONG, M. A Handbook of Human Resources Management Practise, 8th edition, (in Czech.) Praha: Grada, 2002. 777p, ISBN 80-247-0469-2.

Periodicals

2 BONDAROUK, T. V., RUEL, H. J. M. Electronic Human Resource Management: challenges in the digital era. The International Journal of Human Resource Management Vol. 20, No. 3, March 2009, 505P.ISSN 0958-5192.

3 LEE, CH, S., LEE, CH. H., Effects of HRM practices on it usage. Journal of Computer Information Systems. Winter 2009.

4 POWELL, T. C., DENT-MICALLEF, A. "Information Technology as Competitive Advantage: The Role of Human Business, and Technology Resources," Strategic Management Journal. 1997.

5 DeSANCTIS, G. Human Resource Information Systems: A Current Assessment. MIS. Quarterly, 10. 1986.

Other Reports and Documents

6 HELFEN, M., KRUGER, L. Information Technology, New Organizational Concepts and Employee Participation - Will Unionism Survive?. 202p. 0-7803-7824-0/02.

7 SAMENI, M.K., KHOSHALHAN, F. Analysis of Human Resource Development for Information Technology and E-Commerce in Iran. IT Department, Faculty of Industrial Engineering, K.N. Toosi University of Technology, Tehran, Iran. 2006.

8 ZAKARIA, N., YUSOF, S.A.M. The role of human and organizational culture in the context of technological change. School of Information Studies, Syracuse University, Syracuse, USA. 2001. 0-7803-7260-3.

5

Toward Internationalization of Higher Education: Emerging Educational Markets

Dr. Anita Trnavcevic[1], Dr. Nada Trunk Sirca[2]
and Mag. Vinko Logaj[3]

[1]*University of Primorska, Slovenia*

[2]*University of Primorska, Slovenia*

[3]*National School for Leadership in Education, Slovenia*

5.1 Abstract

The purpose of the paper is to present and discuss the findings of the needs analysis in SE Europe and to reflect upon them through 'cultural diversity perspective' and marketization point of view. University of Primorska, Faculty of Management Koper, in accordance with the flow of internationalization, has developed a post-graduate study programme in educational management and leadership that could best serve the needs of the region and also beyond 'regional borders'. As a part of marketing approach to development and dissemination of new programmes, needs analysis in the area of educational management and leadership was carried out.

Keywords: Higher education, Internationalisation, Marketization, Commodification of education.

5.2 Introduction

In the area of education, the last decade has been coloured with major changes in educational policies which are embraced in processes such as decentralization, deregulation, de-concentration and lump-sum financing. These processes are labelled in the current theoretical discussions as marketization of education' (Kenway and Bullen 2001; Dehli 1996), which, on the other hand, is not the same as privatization in education. Marketization of education is closely related to globalization, internationalization of education and 'audit culture'.

If 'globalization' can be understood as the most (mis)used, almost mysterious and rarely well defined word (Beck 2003: 37) then internationalization, especially in higher education, is a 'more operational concept' than globalization. Teichler (2004) argues that internationalization in higher education means increased activities beyond the borders of nation-states and is closely related to 'physical' mobility, exchange, cooperation between higher education institutions, 'harmonization' of higher education (Bologna processes) and also to the emergence of educational markets. Higher education institutions focus their efforts on the international recognition and comparability, and design programmes that can best meet the needs of 'international' students (see www.eurydice.org; Teichler 2004). Knowledge, embraced in educational programmes is transformed into commodity, and driven by 'demand-supply' relationship.

From critical, sociology-based perspective, commodification of higher education should be subjected to scrutinized analysis and discussions about consequences on the production, dissemination and use of knowledge as well as with respect to the role it plays in the globalised world of managerialism and neo-liberal politics (see Ball 2004; Beckman and Cooper 2004; Apple 2004; Baldwin and James 2000).

From the management and marketing point of view the essential question is related to audits, needs assessment, design and dissemination of programmes tailored as much as possible to best serve the needs of 'international' customers in order to embrace cultural diversity. Changing nature of education and the purpose of knowledge is not at the heart of 'marketing' discussions.

The purpose of the paper is to present and discuss the findings of the needs analysis in SE Europe and to reflect upon them through 'cultural diversity perspective' and marketization point of view. University of Primorska, Faculty of Management Koper, in accordance with the flow of internationalization, has developed a post-graduate study programme in educational management and leadership that could best serve the needs of the region and also beyond 'regional borders'. As a part of marketing approach to development and dissemination of new programmes, needs analysis in the area of educational management and leadership was carried out.

5.3 Theoretical Background

In the last decade, higher education in Slovenia has been subjected to many changes, some of them associated with the transformation of university programmes, content and structure, in accordance with the Bologna processes. Some changes can be related to major shifts from 'socialist' to market-driven society and recently the emergence and expansion of market fundamentalism, associated with neo-liberal trends and discussions. Current trends in the Slovenian higher education can be discussed in the light of 'market-based' society. On the other hand, even a stronger standpoint can be taken by saying that the first traces of market fundamentalism can be noticed in the development of higher education. Somers and Block (2005) argue that "over the past twenty years, 'market fundamentalism' has moved from the margins of debate to become dominant policy perspective across the global economy" (p. 260). In their view, the concept of market fundamentalism can by no means be mixed with "the complex mix of policies pursued by governments in actually existing market societies" (p. 261), Somers and Block (2005) point out that market fundamentalism is more extreme than the nuanced arguments of mainstream economists; it emphasizes and subordinates, and in a way, suppresses society as whole to a system of self-regulating markets. Higher education institutions are, as other educational institutions, rigid and slow in introducing even necessary changes related to better response to movements in other areas of social life. However, this 'rigidness' can also be seen as positive when it critically resists the 'new managerialism' in higher education institutions as well as to market fundamentalism.

Relying upon 'supply-demand' exchange in the market has some implications, one of them being the commodification of education. Miller (2003) in his discussion on commodification of higher education in the USA argues, that commodification of education is essentially related to understanding students as consumers, making rational choices. "/../ we can see a tendency across the entire degree-granting sector of transferring the cost of running schools away from governments and towards students, who are regarded more and more as consumers who must manage their own lives, and invest in their own human capital /.../ (p. 901). In his view, there is an important consequence which has redesigned and restructured academic institutions. "Academic institutions have come to resemble their entities they serve now; colleges have been transformed into big businesses. /.../. The mimetic managerial fallacy also leads to more and more forms of surveillance from outside" (p. 902).

He argues that in the US they have noticed "ever-increasing performance-based evaluations of teaching conducted at the departmental and Decanal level, rather than in terms of the standard of an overall school" (p. 897), which are linked to budgeting. Such policies and politics of higher education lead to realignment of power and also to transformation of 'knowledge'. Education, as Lyotard (1984) argues, has become a commodity and so has knowledge. Roberts (1998) claims that "the philosophy of 'user pays', routinely cited as a justification for charges in a whole range of public service areas, has become the order of the day" (p. 7).

Slovenian higher education has been moving in line with current trends. Internationalization, grounded in standardization, performativity, enhanced 'audit culture', and 'quality assurance', technically operated by credits and their transfer, and comparable in terms of programmes (structure and content) has coloured changes in higher education. Alike universities around the world Slovenian universities are also seeking new markets and try to expand their network beyond national borders.

In the SEE, Slovenia has been well positioned for students who are seeking programmes in educational management, leadership and administration. In order to best meet the needs in the region, a survey was conducted. The questionnaire was electronically accessible on the faculty's web page and e-mails were sent to 24 e-mail addresses in the region, all to persons who had been in contact with the Faculty previously and have hold positions in the NGO's, Ministries of Education and other professional institutions. These persons were also asked to disseminate the information about the survey to their colleagues who work in education. The results of the needs analysis were meant to provide informed 'ground' for programme design.

5.4 The Analysis

Despite personal encouragement to a group of approximately 30 participants from the region who attended the meeting in Macedonia in May 2005, and two e-mail requests to 24 e-mail addresses, only 16 questionnaires were filled in. This section, therefore, cannot 'provide' data that could be used as substantial and generalized contribution to a better understanding of the needs in the region and hence to tailor programmes toward these need. Rather, we use the data and take the concept, developed by Sorners and Block (2005), labelled 'market embeddedness' to be suggesting that market based development of higher education ignores cultural diversity and actually leads to the 'sameness' in terms of programmes' structure, content and modes of delivery. Firstly, we

Table 1 Frequency of the Estimation of Courses Needed for the Work

Courses	5	4	3	2	1
Management Processes	14	1	1	0	0
Organisation Theory and Managing External Relations	8	7	1	0	0
Managing People	14	0	2	0	0
Business Communication	7	1	4	4	0
Communication in Educational Practice	12	3	0	1	0
Leadership in Educational Organisations	11	4	0	2	0
Classroom Management	6	4	2	2	1
Curriculum Management	8	4	1	2	1
Managing Quality	12	4	0	0	0
Managing Continuing Professional Development	10	4	2	0	0
Managing Teacher Performance	6	4	4	2	0
Marketing in Educational Organisations	7	2	5	1	1
Effectiveness and Improvement of Educational Organisations	11	2	3	1	0
Evaluation of Educational Organisations	11	0	1	3	1
Strategic Management and the Management of Change	13	2	1	0	0

present the data gathered by a survey and secondly, we discuss them through a cultural diversity perspective in relation to market embeddedness.

5.4.1 About the Needed Courses

16 respondents in the survey, 10 female and 6 male, are from 12 different countries from SEE region, with working experience between 2 to 21 years. They work for HE institution (4), Ministry of Education (1), NGO (5), industry (1), public schools (4) and private school (1). On the scale from 5 to 1 (5 highly needed and 1 not needed at all) they were asked to estimate how the following courses are needed for the work carried out by themselves and their colleagues at work.

Data show that the modules/content of a programme entitled *Post-graduate programme in the area of educational management, administration and policy* highly correspond to the modules. Identified in a study of Erculj et al. (2005) where the comparison of training programmes for educational leaders in 8 European countries was done. Management processes, managing people, communication in educational practice, leadership in educational institutions, managing continuous professional development, managing quality, effectiveness and improvement of educational institutions, evaluation of educational institutions and strategic management are the modules that at least 10 out of 16 respondents assessed as the most needed in the programme. These modules are in tune with the policy trends and developments in Europe and worldwide. Quality, effectiveness and improvement, evaluation and strategic management are, in the current 'knowledge rhetoric' the imperatives of educational institutions and national education policies. In a way they can be seen as part of the market embeddedness discourse, where "ideas, public narratives, and explanatory systems by which states, societies, and political cultures construct, transform, explain, and normalize market processes. No less than all the familiar mechanisms by which markets are shaped, regulated, and organized, so too, they are always ideationally embedded by one or another competing knowledge regime" (Somers and Block, 2005: 264). This market embededdness has almost taken a trans-national' character which is needed for the programmes being accepted internationally and is reflected in explanatory system of the audit culture. Markets, although claimed to be 'free markets' do not lead toward deregulation of educational systems but, paradoxically, toward re-regulation and increased re-centralization within different paradigm. The promise of 'free markets has been critically reflected by McCowan (2004) who argues that education systems worldwide are far from resembling the free markets in which commodities are traded. In his view, even 'market proponents' tend to advocate a system in which some state intervention is necessary. Apple (2004: 617) points to enhanced power of the State as there has to be "the constant production of evidence that you are doing things efficiently and in the correct way".

Needs surveys play an essential role in the estimation of correctness and efficiency when an institution is launching a new programme, especially if the programme is meant to be stretched beyond the national border. These surveys also succumb to the rhetoric of 'addressing cultural diversity', where data gathered by such surveys can be used for a claim that programmes are internationally relevant, needed and comparable. Our data were gathered by Likert scale, there is a specific 'limitation' or a framework, implicitly imposed to respondents, to judge and estimate their needs within internationally recognized and comparable programmes rather than to identify their specific needs emerging from different cultural and educational traditions. Open question about their suggestions partially reduced the 'limitation'. 5 respondents suggested some topics and 1 suggested the area (managing quality in health care). Interestingly, these suggestions can also be seen in the light of current education rhetoric, because they focused on ICT, knowledge management and intercultural communication - the latest trends in education policies worldwide.

Table 2   Frequency of the Estimated Relevance of the Programme for Target Groups

Target Groups	5	4	3	2	1
Teachers in schools	10	3	1	2	0
Principals (head-teachers)	14	1	1	0	0
Policy makers in the educational area at the local level	10	3	2	1	0
Policy makers in the educational area at the municipality level	10	3	2	1	0
Policy makers in the educational area at the regional level	11	4	1	0	0
Policy makers in the educational area at the national level	15	0	1	0	0
Policy makers from non-profit organizations	10	5	1	0	0
Employees in non-profit organizations	8	6	2	0	0

5.4.2 About the Relevance of the Programme

In order to find out relevance of the programme for different target groups, respondents were asked to mark on the scale 5 to 1 (5 highly relevant and 1 not relevant at all) the level of relevance of the programme.

At least 10 out of 16 respondents found the programme highly relevant for teachers in schools, principals, policy makers in the educational area at the local level, at the municipality level, at the regional level and at the national level and policy makers from NGO's. 'Only' 8 out of 16 respondents found the programme highly relevant for employees in those institutions and in NGO's.

Educational management and administration (leadership) post-graduate programmes are focused on these groups of participants (see, for example, MMU and University of Antwerp). The question, however is, how these programmes reflect cultural and educational 'national' traditions and how 'universal' they are. In the international context, by identification of a large group of potential customers, cultural diversity is seemingly addressed because target groups are not 'the same' by the position they hold in an institution, but also by their cultural background.

Taking a stance that marketing addresses cultural diversity (see Kotler 2001) and through segmentation requires 'knowing the customer' and responding to specific needs of a segment, then needs analysis should go beyond general facts about 'content needed' and 'content relevance'. However, in-depth market analysis would require profound data gathering from local environments. A survey, electronically accessible and based on the assumption of ICT breaking the national borders, failed to provide extended data. Also, the questionnaire was only a 'slip' into the surface of needs. Therefore, data cannot be built into a solid foundation for programme design that would guarantee success in the region.

5.4.3 Modes of delivery and ICT

15 out of 16 respondents found the online delivery of the programme accept-able. Their answers indicate that they have access to ICT. We can only speculate that they find themselves comfortable with online studies. From the provider's side this data indicates that online could be the 'right' way for provision of educational programme, especially when the mode of delivery, being summer school 'on-site', and online courses during the academic year, are combined. The answers also indicate that the combination of modes of delivery, acceptance of ICT and low opportunities to afford the programme if no scholarships are ensured, leads to 'addressing' diversity and different needs of students in the best possible way. Programmes might be seen more 'marketable' if they are ICT based, and they can address diversity by loosing 'spatial' limitation, but it would also be worth considering that ICT 'unifies' within 'cyberspace'. We are taking Kenway and Buullen (2001; 141) critical argument about style that "allows people and schools to imagine themselves differently, providing an opportunity to define and redefine themselves. To quote Barthes (1975), It can be 'a dream of identity' " to be suggesting that ICT enables to define and redefine people, which can be 'a dream of identity', but they also construct marketable power of consumers in relation to education institution. In 'physical' world this power is very much limited by financial sources and resources.

5.5 Concluding Discussion

We started the survey with the purpose to gather data and hence to build the new programme on solid data-based grounds. We were concerned with cultural diversity issues and the idea that needs analysis as part of marketing can be the best 'tool' to find out the needs that would reflect cultural background of potential students. As the study evolved some other issues were raised, therefore the concluding discussion can be divided into two parts - needs analysis, data and conclusions and reflections upon this data through market embeddedness and market fundamentalism point of view.

5.5.1 Discussion About the Survey

The questionnaire consisted of 14 close-ended and open-ended questions. It was electronically accessible at the Faculty's web page. Despite kind requests to 24 colleagues to disseminate the information about the survey among their colleagues and despite the encouragement of a group attending a seminar in Macedonia and two additional e-mail requests, only 16 respondents filled in the questionnaire. The respondents were from 12 different SEE countries. They worked for different educational institutions and NGO's. They also differed with regard to their working experience. In their views the courses proposed in the new educational management and administration (leadership) programme were needed and relevant for different groups of employees in educational institutions and NGO's. They found online course delivery acceptable and were also keen on summer school as well as on other modes of delivery. However, they lacked personal financial resources in order to be able to enrol in the programme. Therefore, some financial support was expected in the form of scholarships that could cover the accommodation during the 'face-to face' provision. The data could not be generalized and could not provide the ground on which the faculty management could make decisions.

One might argue that sending e-mails to 24 respondents and receiving 16 questionnaires, represents a high response rate. However, these 24 persons were not meant to be the whole sample. They were asked to spread the information about the survey and hence to acquire more respondents, It seems that the survey 'failed' in terms of the size of sample although low response rate has been reported by our Master students and others, who conducted surveys. The question is, however, if a survey is needed when programmes are internationally comparable in terms of the content, structure and also mode of delivery and how can a needs analysis in the area of educational programmes granting degree contribute to changes in the programme.

5.5.2 Reflections

Addressing cultural diversity has become an imperative for educational systems worldwide. Education, although nationally 'bounded', has, especially in the area of higher education, crossed national borders and became an international 'business'. Education, transformed into commodity in the marketplace, has to, therefore, address cultural diversity and the needs of 'international' customers. There are many ways to address diversity and one of them, emerging from 'market embeddedness' and traced 'market fundamentalism' is also related to market and managerialism-oriented approach of higher education institutions to 'run their business'. Part of this 'business' is also to expand educational markets beyond national borders. When going 'beyond', cultural diversity needs to be, addressed in order to be competitive. However, addressing diversity requires more than embellishments' in educational programmes, which could also be labeled as 'style' packaging which could create a dream of identity. Diversity is closely associated with differences, specifics and local traditions rather than constituting specific 'sameness' in educational programmes, grounded in 'audit culture' and comparability. The question we raise is related to educational markets and the requirement to be different yet within the 'audit culture' and cornmodification of education this notion of 'different' is embedded within 'the sameness'.

Our purpose is not to provide a nostalgic view of education being isolated and closed within universities and research institutes. Rather we aimed at raising critical perspective to current marketization and commodification of education flows and to set the ground for further discussion. The survey reflects the need of the faculty to be internationally comparable and competitive, while data which are rather 'thin' and not generalizable reflect specific 'market embeddedness' especially when topics, courses and target groups are discussed. Respondents did not suggest 'specific' topics that would not be included in any education management and administration Master programme, perhaps the only 'new' piece of information was about the lack of resources to afford studies. However, this information too was assumed even before the survey was disseminated. What, then, was gained?

References

1 Apple, W. M. 2004. Schooling, Markets, and an Audit Culture. Educational Policy 18(4):614–621.
2 Baldwin, G. & James, R. 2000. The market in Australian higher education and the concept of student as informed consumer. Journal of Higher Education Policy and Management 22(2), 139–148.
3 Ball, S. 2004. Education for sale! The commodification of everything? King's Annual Education Lecture 2004. University of London. Institute of Education.
4 Beck, U. 2003. Kaj je globalizacija? Zmote globalizma - odgovori na globaliiacijo. Ljubljana: Kit.
5 Beckmann, A. & Cooper, C. 2004. 'Globalization', the new managerialism and education: Rethinking the purpose of education in Britain.

The Journal for Critical Educational Policy Studies, 2 (2)
http://wwwjcepsxom/?pageID=articIe & articleID=31

6 Dehli, K. 1996. Between 'Market' and 'State'? Engendering Education Change in 1990s. Discourse Studies in the Cultural Politics of Education, (17)3: 363–376.

7 Erculj, J. et al (2005). ESIST: Evaluation far emproving school leaders training programmes. - Ljubljana: Sola za ravnatelje, 2005

8 Kenway, J. and Bullen, E, 2001. Consuming Children: Education Entertainment- Advertising. Open University Press: Buckingham.

9 Kotler, P. 2001. A framework for marketing management. Upper Saddle River, N.J: Prentice Hall.

10 Lyotard, J-L. 1984. The postmodern condition: A report on knowledge. Mihneapols: University of Minnesota Press.

11 McCowan, T. 2004. Toolay's seven virtues and the profit incentive in higher education. Journal for Critical Education Policy Studies, 2(2), www.jceps.com.

12 Miller, T. 2003. Governmental or, commodification? US Higher Education, CulturalStudies, 17(6), 897–904.

13 Roberts, P. 1998. Rereading Lyotard: Knowledge, commodification and higher education, Electronic Journal of sociology. www.sociology. org/content/vol003.003/roberts.htmL Accessed 31.5.2005.

14 Somers, M. R. and Block, F. 2005. From poverty to perversity: Ideas, Markets, and Institutions over 200 Years of Welfare Debate. American Sociological Review, 70(April) 260–287.

15 Teichler, Ul. 2004. The Changing Debate on Internationalisation of Higher Education. Higher Education 48: 5–26. http://www.eurydice.org+

6

The Development of Human Resource Management from a Historical Perspective and its Implications for the Human Resource Manager

Franklyn Chukwunonso

Department of Information Technology,
Federal University of Technology, Yola, Nigeria

6.1 Abstract

This paper introduces the development of Human Resource Management (HRM) from a historical Perspective and explains the debate between HRM and personnel management. Thus, the paper identifies the historical developments and their impacts on HRM, outlines the development and functions of HRM, explains the differences between HRM and Personnel Management, evaluates 'hard' and 'soft' approaches to HRM, illustrates how diversity is an issue in Human Relations (HR) practice and finally considers HRM as an international issue. It concludes with a discussion about 'hard' and 'soft' models of HRM and its implications for the human resource manager.

6.2 Introduction

The term "human resource management" has been commonly used for about the last ten to fifteen years. Prior to that, the field was generally known as "personnel administration." The name change is not merely cosmetics.

Personnel administration, which emerged as a clearly defined field by the 1920s (at least in the US), was largely concerned the technical aspects of hiring, evaluating, training, and compensating employees and was very much of "staff" function in most organizations. The field did not normally focus on the relationship of disparate employment practices on overall organizational performance or on the systematic relationships among such practices. The field also lacked a unifying paradigm.

HRM developed in response to the substantial increase in competitive pressures American business organizations began experiencing by the late 1970s as a result of such factors as globalization, deregulation, and rapid technological change. These pressures gave rise to an enhanced concern on the part of firms to engage in strategic planning-a process of anticipating future changes in the environment conditions (the nature as well as level of the market) and aligning the various components of the organization in such a way as to promote organizational effectiveness.

Human resource management (HRM), also called personnel management, consists of all the activities undertaken by an enterprise to ensure the effective utilization of employees toward the attainment of individual, group, and organizational goals. An organization's HRM function focuses on the people side of management. It consists of practices that help the organization to deal effectively with its people during the various phases of the employment cycle, including pre-hire, staffing, and post-hire. The pre-hire phase involves planning practices. The organization must decide what types of job openings will exist in the upcoming period and determine the necessary. Qualifications for performing these jobs. During the hire phase, the organization selects its employees. Selection practices include recruiting applicants, assessing their qualifications, and ultimately selecting those who are deemed to be the most qualified. In the post-hire phase, the organization develops HRM practices for effectively managing people once they have "come through the door." These practices are designed to maximize the performance and satisfaction levels of employees by providing them with the necessary knowledge and skills to perform their jobs and by creating conditions that will energize, direct, and facilitate employees' efforts toward meeting the organization's objectives.

6.3 The Historical Background of Human Resource Management

Human resource management has changed in name various times throughout history. The name change was mainly due to the change in social and economic activities throughout history.

6.3.1 Industrial Welfare

Industrial welfare was the first form of human resource management (HRM). In 1833 the factories act stated that there should be male factory inspectors. In 1878 legislation was passed to regulate the hours of work for children and women by having a 60 hour week. During this time trade unions started to be formed. In 1868 the 1st trade union conference was held. This was the start of collective bargaining. In 1913 the number of industrial welfare workers had grown so a conference organized by Seebohm Rowntrle was held. The welfare workers association was formed later changed to Chartered Institute of Personnel and Development.

6.3.2 Recruitment and Selection

It all started when Mary Wood was asked to start engaging girls during the 1st world war. In the 1st world war personnel development increased due to government initiatives to encourage the best use of people. In 1916 it became compulsory to have a welfare worker in explosive factories and was encouraged in munitions factories. A lot of work was done in this field by the army forces. The armed forces focused on how to test abilities and IQ along with other research in human factors at work. In 1921 the national institute of psychologists established and published results of studies on selection tests, interviewing techniques and training methods.

6.3.3 Acquisition of Other Personnel Activities

During the 2nd world war the focus was on recruitment and selection and later on training; improving morale and motivation; discipline; health and safety; joint consultation and wage policies. This meant that a personnel department had to be established with trained staff.

6.3.4 Industrial Relations

Consultation between management and the workforce spread during the war. This meant that personnel departments became responsible for its organization and administration. Health and safety and the need for specialists became the focus. The need for specialists to deal with industrial relations was recognized so that the personnel manager became as spokesman for the organization when discussions where held with trade unions/shop stewards. In the 1970's industrial relations was very important. The heated climate during this period reinforced the importance of a specialist role in industrial relations negotiation. The personnel manager had the authority to negotiate deals about pay and other collective issues.

6.3.5 Legislation

In the 1970's employment legislation increased and the personnel function took the role of the specialist advisor ensuring that managers do not violate the law and that cases did not end up in industrial tribunals.

6.3.6 Flexibility and Diversity

In the 1990's a major trend emerged where employers were seeking increasing flexible arrangements in the hours worked by employees due to an increase in number of part-time and temporary contracts and the invention of distance working. The workforce and patterns of work are becoming diverse in which traditional recruitment practices are useless. In the year 2000, growth in the use of internet meant a move to a 24/7 society. This created new jobs in e-commerce while jobs were lost in traditional areas like shops. This meant an increased potential for employees to work from home. Organizations need to think strategically about the issues these developments raise. HRM managers role will change as changes occur.

6.3.7 Information Technology

Some systems where IT helps HRM are: Systems for e-recruitment; On-line short-listing of applicants; Developing training strategies on-line; Psycho-metric training; Payroll systems; Employment data; Recruitment adminis-tration; References; Pre-employment checks. IT helps HR managers offload routine tasks which will give them more time in solving complex tasks. IT also ensures that a greater amount of information is available to make decisions.

6.4 Historical Milestones in HRM Development

Table 1 identifies some of the major milestones in the historical development of HRM. Frederick Taylor, known as the father of scientific management, played a significant role in the development of the personnel function in the early 1900s. In his book. *Shop Management,* Taylor advocated the "scientific" selection and training of workers. He also pioneered incentive systems that

Table 1   Milestones in the Development of Human Resource Management

1890–1910	Fredick Taylor develops his ideas on scientific management. Taylor advocates scientific selection of workers based on qualifications and also argues for incentives- based compensation systmes to motivate employees.
1910–1930	Many companies establish departments devoted to maintaining the welfare of workers. The discipline of industrial psychology begins to develop. Industrial psychology, along with the advent of World War I, leads to advancements in employment testing and selection.
1930–1945	The interpretation of the Hawthorne Studies' begins to have an impact on management thought and practice. Greater emphasis is placed on the social and informal aspects of the workplace affecting worker productivity. Increasing the job satisfaction of workers is cited as a means to increase their productivity.
1945–1965	In the U.S., a tremendous surge in union membership between 1935 and 1950 leads to a greater emphasis on collective bargaining and labour relations within personnel management. Compensation and benefits administration also increase in importance as unions negotiate paid vacations, paid holidays, and insurance coverage.
1965–1985	The Civil Rights movement in the U.S. reaches its apex with passage of the Civil Rights Acts of 1964. The personnel function is dramatically affected by Title VII of the CRA, which prohibits discrimination on the basis of race, color, sex, religion, and national origin. In the years following the passage of the CRA, equal employment opportunity and affirmative action become key human resource management responsibilities.
1985–present	Three trends dramatically impact HRM. The first is the increasing diversity of the labor force, in terms of age, gender, race, and ethnicity. HRM concerns evolve from EEO and affirmative action to "managing diversity." A second trend is the globalization of business and the accompanying technological revolution. These factors have led to dramatic changes in transportation, communication, and labor markets. The third trend, which is related to the first two, is the focus on HRM as a "strategic" function. HRM concerns and concepts must be integrated into the overall strategic planning of the firm in order to cope with rapid change, intense competiton, and pressure for increased efficiency.

rewarded workers for meeting and/or exceeding performance standards. Although Taylor's focus primarily was on optimizing efficiency in manufacturing environments, his principles laid the ground-work for future HRM development. As Taylor was developing his ideas about scientific management, other pioneers were working on applying the principles of psychology to the recruitment, selection, and training of workers. The development of the field of industrial psychology and its application to the workplace came to fruition during World War I, as early vocational and employment-related testing was used to assign military recruits to appropriate functions.

The Hawthorne Studies, which were, conducted in the 1920s and 1930s at Western Electric, sparked an increased emphasis on the social and informal aspects of the workplace. Interpretations of the studies emphasized "human relations" and the link between worker satisfaction and productivity. The passage of the Wagner Act in 1935 contributed to a major increase in the number of unionized workers. In the 1940s and 1950s, collective bargaining led to a tremendous increase in benefits offered to workers. The personnel function evolved to cope with labor relations, collective bargaining, and a more complex compensation and benefits environment. The human relations philosophy and labor relations were the dominant concerns of HRM in the 1940s and 1950s. HRM was revolutionized in the 1960s by passage of Title VII of the Civil Rights Act and other antidiscrimination legislation—as well as presidential executive orders that required many organizations to undertake affirmative action in order to remedy past discriminatory practices. Equal employment opportunity and affirmative action mandates greatly complicated the HRM function, but also enhanced its importance in modern organizations. As discussed more fully in a later section, these responsibilities continue to comprise a major part of the HRM job. Finally, changes in labor force demographics, technology, and globalization since the 1980s have had a major impact on the HRM function. These factors also are discussed in more detail in a later section.

6.5 The Difference Between HRM and Personnel Management

Some experts assert that there is no difference between human resources and personnel management. They state that the two terms can be used interchangeably, with no difference in meaning. In fact, the terms are often used interchangeably in help-wanted ads and job descriptions. For those who recognize a difference between personnel management and human resources, the difference can be described as philosophical. Personnel management is more administrative in nature, dealing with payroll, complying with employment law, and handling related tasks. Human resources, on the other hand, is responsible for managing a workforce as one of the primary resources that contributes to the success of an organization. When a difference between personnel management and human resources is recognized, human resources is described as much broader in scope than personnel management. Human resources is said to incorporate and develop personnel management tasks, while seeking to create and develop teams of workers for the benefit of the organization. A primary goal of human resources is to enable employees to work to a maximum level of efficiency.

Personnel management can include administrative tasks that are both traditional and routine. It can be described as reactive, providing a response to demands and concerns as they are presented. By contrast, human resources involves ongoing strategies to manage and develop an organization's workforce. It is proactive, as it involves the continuous development of functions and policies for the purposes of improving a company's workforce.

Personnel management is often considered an independent function of an organization. Human resource management, on the other hand, tends to be an integral part of overall company function. Personnel management is typically the sole responsibility of an organization's personnel department. With human resources, all of an organization's managers are often involved in some manner, and a chief goal may be to have managers of various departments develop the skills necessary to handle personnel-related tasks.

As far as motivators are concerned, personnel management typically seeks to motivate employees with such things as compensation, bonuses, rewards, and the simplification of work responsibilities. From the personnel management point of view, employee satisfaction provides the motivation necessary to improve job performance. The opposite is true of human resources. Human resource management holds that improved performance leads to employee satisfaction. With human resources, work groups, effective strategies for meeting challenges, and job creativity are seen as the primary motivators.

When looking for a job in personnel management or human resources, it is important to realize that many companies use the terms interchangeably. if you are offered a job as a personnel manager, you may be required to perform the same duties as a human resource manager, and vice versa. In some companies, a distinction is made, but the difference is very subtle.

6.6 HRM Development and Implementation Responsibilities

While most firms have a human resources or personnel department that develops and implements HRM practices, responsibility lies with both HR professionals and line managers. The interplay between managers and HR professionals leads to effective HRM practices. For example, consider performance appraisals. The success of a firm's performance appraisal system depends on the ability of both parties to do their jobs correctly. HR professionals develop the system, while managers provide the actual performance evaluations. The nature of these roles varies from company to company, depending primarily on the size of the organization. This discussion assumes a large company with a sizable HRM department. However, in smaller companies without large HRM departments, line managers must assume an even larger role in effective HRM practices. HR professionals typically assume the following four areas of responsibility: establishing HRM policies and procedures, developing/choosing HRM methods, monitoring/evaluating HRM practices, and advising/assisting managers on HRM-related matters. HR professionals typically decide (subject to upper-management approval) what procedures to follow when implementing an HRM practice. For example, HR professionals may decide that the selection process should include having all candidates (1) complete an application, (2) take an employment test, and then (3) be interviewed by an HR professional and line manager.

Usually the HR professionals develop or choose specific methods to implement a firm's HRM practices. For instance, in selection the HR professional may construct the application blank, develop a structured interview guide, or choose an employment test. HR professionals also must ensure that the firm's HRM practices are properly implemented. This responsibility involves both evaluating and monitoring. For example, HR professionals may evaluate the usefulness of employment tests, the success of training programs, and the cost effectiveness of HRM outcomes such as selection, turnover, and recruiting. They also may monitor records to ensure that performance appraisals have been properly completed. HR professionals also consult with management on an array of HRM-related topics. They may assist by providing managers with formal training programs on topics like selection and the law, how to conduct an employment interview, how to appraise employee job performance, or how to effectively discipline employees. HR professionals also provide assistance by giving line managers advice about specific HRM-related concerns, such as how to deal with problem employees.

Line managers direct employees' day-to-day tasks. From an HRM perspective, line managers are mainly responsible for implementing HRM practices and providing HR professionals with necessary input for developing effective practices. Managers carry out many procedures and methods devised by HR professionals. For instance, line managers:

- Interview job applicants
- Provide orientation, coaching, and on-the-job training
- Provide and communicate job performance ratings
- Recommend salary increases
- Carry out disciplinary procedures
- Investigate accidents
- Settle grievance issues

The development of HRM procedures and methods often requires input from line managers. For example, when conducting a job analysis, HR professionals often seek job information from managers and ask managers to review the final written product. Additionally, when HR professionals determine an organization's training needs, managers often suggest what types of training are needed and who, in particular, needs the training.

6.7 HRM Specialty Areas or Functions of HRM

TRADITIONAL SPECIALTY AREAS

6.7.1 Training/Development

Conducts training needs analysis; designs/conducts/evaluates training programs; develops/implements succession planning programs.

6.7.2 Compensation/Benefits

Develops job descriptions; facilitates job evaluation processes; conducts/ binterprets salary surveys; develops pay structure; designs pay for performance and/or performance improvement programs; administers benefits program.

6.7.3 Employee/Industrial Relations

Helps resolve employee relations problems; develops union avoidance strategies; assists in collective bargaining negotiations; oversees grievance procedures.

6.7.4 Employment/Recruiting

Assists in the HR planning process; develops/purchases HR information systems; develops/updates job descriptions; oversees recruiting function; develops and administers job posting system; conducts employment interviews, reference checks, and employment tests; validates selection procedures; approves employment decisions.

6.7.5　Safety/Health/Wellness

Develops accident prevention strategies; develops legal safety and health policies; implements/promotes EAP and wellness programs; develops AIDS and substance abuse policies.

6.7.6　EEO/Affirmative Action

Develops and administers affirmative action programs; helps resolve EEO disputes; monitors organizational practices with regard to EEO compliance; develops policies for ensuring EEO compliance, such as sexual harassment policies.

6.7.7　HRM Research

Conducts research studies, such as cost-benefit analysis, test validation, program evaluation, and Feasibility studies.

6.8　New HRM Specialty Areas

6.8.1　Work and Family Programs

Develops and administers work and family programs including flextime, alternative work scheduling, dependent-care assistance, telecommuting, and other programs designed to accommodate employee needs; identifies and screen child- or elder-care providers; administers employer's private dependent-care facility; promotes work and family programs to employees.

6.8.2　Cross-Cultural Training

Translate the manners, mores, and business practices of other nations and cultures for American business people. Other cross-cultural trainers work with relocated employees' families, helping them adjust to their new environment.

6.8.3　Managed-Care

As a company's health-care costs continue to escalate, employers are embracing managed-care systems, which require employees to assume some of the costs. Employers hire managed-care managers to negotiate the best options for employees.

6.8.4 Managing Diversity

Develop policies and practices to recruit, promote, and appropriately treat workers of various ages, races, sexes, and physical abilities.

6.9 Contemporary/Diversity Issues

HRM departments within organizations, just as the organizations themselves, do not exist in a vacuum. Events outside of work environments have far-reaching effects on HRM practices, The following paragraphs describe some of these events and indicate how they influence HRM practices.

As mentioned previously, the enactment of federal, state, and local laws regulating work place behavior has changed nearly all HRM practices. Consider, for instance, the impact of antidiscrimination laws on Firms' hiring practices. Prior to the passage of these laws, many firms hired people based on reasons that were not job-related. Today, such practices could result in charges of discrimination. To protect themselves from such charges, employers must conduct their selection practices to satisfy objective standards established by legislation and fine-tuned by the courts. This means they should carefully determine needed job qualifications and choose selection methods that accurately measure those qualifications.

- Social, economic, and technological events also strongly influence HRM practices. These events include:
- An expanding cultural diversity at the work-place
- The emergence of work and family issues
- The growing use of part-time and temporary employees
- An increased emphasis on quality and team-work
- The occurrence of mergers and takeovers
- The occurrence of downsizing and layoffs
- The rapid advancement of technology
- An emphasis continuous quality improvement
- A high rate of workforce illiteracy

These events influence HRM practices in numerous ways. For example:

- Some firms are attempting to accommodate the needs of families by offering benefit options like maternity leave, child care, flextime, and job sharing.

- Some firms are attempting to accommodate the needs of older workers through skill upgrading and training designed to facilitate the acceptance of new techniques.

- Some firms are educating their employees in basic reading, writing, and mathematical skills so that they can keep up with rapidly advancing technologies.

Unions often influence a firm's HRM practices. Unionized companies must adhere to written contracts negotiated between each company and its union. Union contracts regulate many HRM practices, such as discipline, promotion, grievance procedures, and overtime allocations. HRM practices in non-unionized companies may be influenced by the threat of unions. For example, some companies have made their HRM practices more equitable (i.e., they treat their employees more fairly) simply to minimize the likelihood that employees would seek union representation.

Legal, social, and political pressures on organizations to ensure the health and safety of their employees have had great impacts on HRM practices. Organizations respond to these pressures by instituting accident prevention programs and programs designed to ensure the health and mental well-being of their employees, such as wellness and employee assistance programs.

Today's global economy also influences some aspects of HRM. Many firms realize that they must enter foreign markets in order to compete as part of a globally interconnected set of business markets. From an HRM perspective, such organizations must foster the development of more globally-oriented managers: individuals who understand foreign languages and cultures, as well as the dynamics of foreign market places. These firms also must deal with issues related to expatriation, such as relocation costs, selection, compensation, and training.

6.10 Hard and Soft Approaches to HRM

Human resource as defined by Dessler (2004) is the strategy for acquiring, using, improving and preserving, the organisations human resource. It could be well agued that in most cases the human aspect is forgotten in relation to how they manage people, leaving most staff unsatisfied creating a high staff turn over which affects organizational performance. It is therefore an utmost importance that people as opposed to just employees-need to be managed in away that consistent with broad organizational requirement such as quality or efficiency. As in most cases organizational effectiveness depends on there being a tight 'fit' between human resource and business strategies.

Human resource as could be said is all about making business strategies work. It is therefore important that emphasis is placed on how to best match and develop "appropriate" human resource management (HRM) approach/system of managing people in the tourism hospitality and leisure industry (THL). Thus, we would therefore be looking at some of the HRM approaches used such as the Harvard model; hard and soft approach in conjunction with the real world of the THL industry and to determine whether the hard approach is more appropriate. Human resource management (HRM) as described by Kleiman (2000) has a concept with two distinct forms; soft and hard approach, where the soft approach of HRM is associated with human relation and the hard on the other hand sees people as human resource. The Soft HRM is the notion that workers respond better when an organization recognizes their individual needs and addresses them as well as focusing on the overall business objectives. The work of Maslow in stating that humans have a 'hierarchy' of needs, which they will exert considerable energy towards achieving, claims that organizations that recognizes and addresses these needs will have a happier, more fulfilled, more loyal and productive workforce (SHRM Online). As argued by Noe (2006) the way to success is through deep empathy of other people either by observing how to best 'connect' with others in the workplace, and motivate and inspire them as a result. As illustrated by Simon (1960) all of these soft HRM can of course be balanced by hard HRM; the notion that successful organizations are those that best deploy their human resource in the way that they would deploy any other resource.

The Hard HRM on the other hand therefore sees people as human resource. Holding that employees are a resource in the same way as any other business resource and they must therefore be; obtained as cheaply as possible, used sparingly, developed and exploited as much as possible. As indicated by Kleiman (2000) under this model of HRM, control is more concerned with performance system, performance management and tight control over individual activities with the ultimate goal being to secure the competitive advantage of the organization. The hard HRM therefore is primarily concern to promote human resource strategy and align with business strategy. It may also include out sourcing, flexibility, performance management, hence downsizing or work intensification, sees workers as another resource to be exploited and can operate against the interest of workers.

The Harvard model on the other hand as indicated by Lado and Wilson (1994) sees employees as resource, but human where the managers are responsible to make decisions about the organization and employee relation. The employment relation is seen as a blending of business and societal expectations and because it recognizes the role societal outcomes play, it could be argued that the Harvard model provides a useful basis for comparative analysis. The Harvard model also cover the four HRM policy areas which are human resource flows, reward system, employee influence, work system, which leads to the four Cs; competence of employees commitment of employees, congruence of organization/employees goals and cost effectiveness of HRM. As could be agued striving to enhance all four Cs could lead favorable consequences for individual well-being societal well-being and organizational effectiveness either as long-term consequences.

6.11 Conclusion

The penalties for not being correctly staffed are costly. Planning staff levels requires that an assessment of present and future needs of the organization be compared with present resources and future predicted resources. Appropriate steps should then be planned to bring demand and supply into balance. The central aim of modern human resource management is to enhance the effective use, involvement and contribution of employees throughout the organization. This, clearly, requires a great deal of information accretion, classification and statistical analysis as subsidiary aspect of personnel management. What future demands will be is only influenced in part by the forecast of the human resource manager, whose main task may well be to scrutinize and modify the crude predictions of other managers.

References

1 Dessler, Gary. Human Resource Management. 10th ed. Englewood Cliffs, NJ: Pearson/Prentice-Hall, 2004.
2 Kleiman, Lawrence S. Human Resource Management: A Managerial Tool for Competitive Advantage. Cincinnati: South-Western College Publishing, 2000.
3 Lado, AA, and M. C. Wilson. "Human Resource Systems and Sustained Competitive Advantage: A Competency-Based Perspective," Academy of Management Review 19, no. 4 (1994): 699–727.

4 Noe, Raymond A., et al. Human Resource Management: Gaining a Competitive Advantage. 5th ed. Boston: McGraw-Hill, 2006.

5 SHRM Online, Society for Human Resource Management. Available from http://www.shrm.org

6 Simon, H. A., The New Science of Management Decision, New York, NY: Harper and Row, 1960.

7 Swanson, E. B. and M. J. Culnan, "Document-Based Systems for Management Planning and Control: A Classification, Survey, and Assessment", MIS Quarterly, 2, 4, Dec, 1978, 31–46.

8 Urban, G. L., "SPRINTER: A Tool for New Products Decision Makers", Industrial Management Review, 8, 2, Spring 1967, 43–54.

7

Developing Strategic International Human Resource Management: Prescriptions for MNC Success

Mary Ann Von Glinow and John milliman

Centre for effective Organization, Marshall School of Business, University of Southern California, Los Angeles, CA, USA

7.1 Abstract

Many U.S. multinational corporations (MNCs) have experienced difficulties in their overseas operations, due in part to ineffective international human resource management (IHRM) practices. This paper uses a product life cycle (PLC) approach to develop a two-step contingency model of the strategic and operational levels of MNCs. Effective IHRM practices which are tailored to the specific characteristics and needs of the MNC and its environment are discussed. Propositions on the effectiveness of IHRM practices are developed based on the PLC and contingency model. We believe that the adoption of this contingency approach can be critical in enabling U.S. Firms to create the level of cross-cultural managerial effectiveness needed at each phase of production and ultimately create a global vision which will be necessary to compete in many highly competitive and rapidly changing industries.

7.2 Introduction

The United States' International human resources management (IHRM) research (Harvey, 1981; Tung, 1981) and business practices have not kept pace with the rapid changes occurring in an increasingly competitive global economy during the past two decades (Adler & Ghadar, 1989; Evans, 1987). U.S. firms have relatively high failure rates of expatriate assignments (Conway, 1984) and far fewer formal selection and training programs than either

Japanese or European companies (Mendenhall & Oddou, 1985; Tung, 1988). Evans (1987) cogently noted that a review of the literature from the late 1960's reveals virtually no change in the IHRM strategies of MNCs. These trends remain despite the fact that HRM practices have increasingly been viewed as a critical factor in the success of the both the domestic industries in Japan (Davis, Kerr, & Von Glinow, 1987) and the U.S. (Peters & Waterman, 1982) as well as in International operations (e.g. Adler & Ghadar, 1989; Pucik, 1984; Rugman, 1988; Tung, 1988). Ironically, It is much easier to prescribe what organizations should do then it is for firms to implement effective IHRM practices within the framework of their global strategic thinking (Rugman, 1988). In order to facilitate more effective IHRM practices this paper has three major objectives. First, to develop a two-step contingency model involving both the strategic and operational aspects of how U.S. multinational companies (MNCs) can develop and implement effective IHRM practices. Second, in order to make this contingency model sensitive to the firm's environment, a product life cycle (PLC) approach is employed to show how IHRM practices and cross-cultural interactions will vary at different times with the development of its products and operations. Third, to further develop the practical value of this contingency model a number of IHRM practices are suggested. This contingency model is a first attempt at modeling strategic and operational differences across a MNC's PLC and serves as the basis for some of the critical research propositions offered later. Further, this model should assist MNCs in developing IHRM practices which are sensitive to their particular needs, but also ultimately result in the creation of a cadre of managers with a global vision to guide the firm in an increasingly competitive and rapidly changing world economy.

7.3 Overview of IHRM Contingency Model

To accomplish an effective international orientation, research and anecdotal data suggest that the MNC must take two major discrete, yet connected steps. These steps are predicated on an articulated global strategy of the MNC. First, the MNC's global strategy must be carefully translated into specific IHRM objectives and goals (Miller, Beechler, Bhatt, & Nath, 1986; Pucik, 1984). Second, on an operational level specific IHRM practices and decisions must be developed (Tung, 1988). Figure 1 illustrates a summary overview of this process, which we model within a contingency framework.

This model illustrates how IHRM decisions can be effectively formulated and implemented. As depicted in Figure 1, the first stage of this model

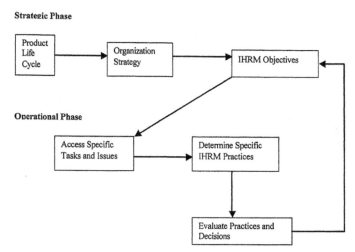

Figure 1 Summary of Two-Step Contingency Model for Developing Effective IHRM Practices

represents a product life cycle (PLC) approach. This PLC model, first espoused by Vernon (1966) has been regularly used to describe how MNCs have evolved since World War II. Originally used to describe trade and investment, MNCs were said to evolve along predictable lines depending upon the particular products they produced. Vernon (1966) portrayed the international PLC as three distinct phases: high tech, growth and internationalization, and maturity. These three phases generally took between 15 to 20 years to complete. Adler & Ghadar (1989) have shown how an accelerated PLC approach can be useful in understanding the evolution of the MNC. Their contention is that such an approach can be helpful on shedding light on the nature of firm's operations, environment, and its strategy. We expand on the early work of Vernon (1966) and Adler and Ghadar (1989) in this manuscript by refining the accelerated PLC approach at the operational level of the MNC. As such, we suggest that the business objectives and strategy of the organization must then be subsequently translated into strategic IHRM objectives (Miller, Beechler, Bhatt, & Nath, 1986). The IHRM strategic goals may then be used to inform the MNC's managers and Personnel Department on criteria for making decisions at an operational level. The specific characteristics of the task or project in conjunction with the criteria for decision making dictate specific IHRM practices. These IHRM practices and decisions should be evaluated and the feedback used to alter strategic IHRM objectives. This model will now be developed more fully by discussing the PLC in the strategic phase.

7.4 Strategic Phase of Contingency Model

7.4.1 The PLC approach and organizational strategy

The strategic phase of the contingency model involves how the MNC's PLC affects its overall business strategy and the relationship between business strategy and IHRM objectives. The critical point is that IHRM objectives vary across different PLC phases (Adler & Ghadar, 1989; Evans, 1987). According to Adler and Ghadar (1989) an important step to developing effective managerial practices is linking cross-cultural and IHRM practices to the specific strategies and contextual factors of the MNC: "Discussions concerning the influence of culture on strategic efficacy remain time-lagged, disconnected from other corporate realities. We continue to ask if culture impacts organizational functioning rather than the more relevant when, or under what conditions, does it do so." (p. 3)

Thus, the PLC approach enables us to better understand what critical factors affect IHRM decisions and practices. In essence, the PLC frames the key characteristics and conditions which MNCs typically face over time in both their internal and external environments. The four phase international PLC approach offered as a synthesis of earlier literature by Adler and Ghadar (1989) is briefly summarized below:

7.4.2 Phase 1: Product orientation

In the first phase the firm develops and maintains essentially a domestic focus which begins by servicing its own internal markets. Slowly it begins to sell its products to foreign nations. The company usually has an ethnocentric and short-term focus and its major emphasis is on product development and research and development (R & D). The company may develop an international division which handles operations for all overseas businesses (Pucik, 1984).

7.4.3 Phase 2: Market orientation

The firm begins to evolve from a domestic orientation to an international orientation and seeks to establish its operations abroad. Threatened with greater competition and in turn seeking to expand their foreign markets, the firm's primary focus now becomes efficiency of production Initially the firm may develop overseas sales offices and assume a polycentric view on crosscultural issues. At some point the organization is likely to develop International product divisions (Adler & Ghadar, 1989; Pucik, 1984).

7.4.4 Phase 3: Price orientation

With overseas business now representing a significant if not dominant portion of the MNC's revenues there is greater focus on price competition and production costs. This phase is typically marked by strong multinational competition and competitive advantage is gained by placing production in countries where the lowest costs can be obtained. The firm may establish global lines of business (Adler & Ghadar, 1989) or a global matrix structure (Pucik, 1989) which coordinates and controls foreign operations in the firm's overall (global) operations.

7.4.5 Phase 4: Price competition

In an era of increasing internal competition and rapid change the MNC faces two simultaneous challenges: it must be highly sensitive to the host country and culture (differentiation), and it must manage the overseas operations within the firm's overall global strategy and objectives (integration) (Von Glinow & Mohrman, in press). This phase can involve the development of a complex set of new organizational forms representing multicentric interests such as cooperative alliances, consortia, networks, joint ventures, and other non-traditional structural forms (Galbraith, in press). To compete more effectively within these new global alliances, existing IHRM practices need to be redefined (Lorange, 1986), and newer more innovative practices must be devised.

Numerous researchers have observed that today's greater competition and faster changes make the PLC phases much shorter than they were before (Riggs, 1983). In the post-World War II era, as was mentioned, it was not uncommon for a firm to go from the first phase, product orientation, through to the price orientation of phase 3 in 15 to 20 years. At present, it is not unusual for firms to move from the initial phase to phase 3 or 4 in three to four years (Adler & Ghadar, 1989). And in the case of high technology firms that time period is severely reduced (Von Glinow, 1988).

These rapid PLC changes have two major consequences for U.S. IHRM activities. First, U.S. IHRM practices often lag a phase to each PLC stage. For example, phase 1's IHRM practices focusing on technical work skills and product development are often continued during phase 2's marketing orientation or phase 3's price orientation (Adler & Ghadar, 1989). Second, many MNCs simply have not developed the advanced IHRM systems and processes necessary to meet the demands of the firm and its environment in PLC phases 3 and 4. Accordingly, the relationship between the organization's PLC and business strategy to IHRM objectives must be addressed.

7.5 Strategic IHRM Characteristics

A critical aspect in the success of the MNC is the translation of its business objectives into strategic IHRM objectives and practices (Miller, Beechler, Bhatt, h Nath, 1986). One limitation of the extant literature noted earlier is that IHRM lacks currency. While Lorange (1986) notes that newer and more innovative practices are required, the literature is imprecise in describing these new innovative practices. We assert that much of the research on the nature of IHRM practices revolves around the following four core issues: timing, cost versus development integration, and differentiation.

7.5.1 Timing

Timing is one of the most important of the core elements and underlies to some extent, the other fundamental issues. Timing involves whether the MNC has a short-term or long-term orientation in its overseas business strategy and IHRM objectives. A short-term orientation requires that the company implement its IHRM practices more quickly. A longer term orientation involves more formal commitment to expatriates and overseas operations and ultimately can result in more effective overseas operations (Tung, 1988). Thus, a careful assessment of the short-term and long-term perspectives should be made, to facilitate the establishment of later priorities.

7.5.2 Cost versus development

Similar to the timing orientation, the emphasis on cost versus development indicates whether the MNCs priority is to incur Initial lower operating costs in selection and training or to focus on the longer-term development of the firm's overseas operations and the career paths of its expatriates.

7.5.3 Integration

Integration involves the degree to which the organization seeks to achieve a corporate global orientation with strong linkages between the corporate and overseas offices in two ways. One is using tight controls exercised by the corporate office. A second is control created through informal organizational culture via international staffing practices (Adler& Ghadar, 1989; Doz & Prahalad, 1981; Edstrom & Galbraith, 1977; Jaeger, 1982). Ondrack's (1985) detailed study of four European and U.S. MNCs supported the contentions of these authors that international staffing practices can provide them with a cadre of managers who can implement an informal control system for diverse operations.

7.5.4 Differentiation

The firm's major IHRM objectives should also dictate differentiation or the extent to which the company will overtly seek to understand the host country's culture. To achieve differentiation the MNC must continuously involve a number of host country nationals and a three-way congruence between management, societal values, and organization structure is needed to make cross-cultural management practices effective (Davis, Kerr, & Von Glinow, 1987). An experienced network of parent country and host country managers can also facilitate communication to and from overseas units to corporate headquarters (Ondrack, 1985). The next section illustrates how each of these major strategic IHRM criteria varies with the PLC of the international firm. Table 1 depicts these relationships.

7.5.5 Phase 1 Product orientation

In PLC phase 1, the product orientation, the MNC has a short-term view on its IHRM practices as it is focusing on domestic R & D and just beginning overseas sales. Most selection and IHRM practices during this phase are conducted on an ad hoc basis; expatriates are usually selected for their technical work skills and usually they possess little intercultural training (Adler & Ghadar, 1989; Pucik, 1984; Tung, 1988). Similarly, MNCs will usually value cost reduction over long-term development and will not have a strong need for corporate management integration or differentiation since they are just beginning overseas operations.

7.5.6 Phase 2 Market orientation

Firms in the market orientation phase are in the process of establishing their overseas operations and are beginning to think in more global terms. For this reason MNCs in phase 2 are still likely to value short term cost savings over longer term career development of expatriates, although in general they will begin to place a greater emphasis on their overseas managers. Similarly, MNCs probably do not see a strong need for integration since the International

Table 1   Ihrm Strategic Issues and Product Life Cycle (Plc) Phase

PLC Phase/ Business Objectives	Short Versus Long Term	Cost Versus Development	Integration Need for Global Strategy	Differentiation Cultural Sensitivity Required
I PRODUCT ORIENTATION	Short term	Cost	Low	Low
1. Product Development				
2. R & D				
II MARKET ORIENTATION	Medium	Cost	Low	Medium
1. Efficiency				
2. Expand Markets				
III PRICE COMPETITION	Long	Cost	High	Low-medium
1. Price Strategy				
2. Integrate Global Operations				
IV PRICE COMPETITION	Long	Development	High	High
1. Price & Quality				
2. Develop MNC and Host Country Cultures				

divisions are just being established. Nonetheless with more advanced operations the firm must be much more sensitive to the needs of the host country and will seek to develop some level of intercultural sensitivity and language skills of its expatriate managers. The cultural orientation of the firm is likely to change from ethnocentric in first phase to either polycentric or region centric in phase 2 (Adler & Ghadar, 1989).

7.5.7 Phase 3 Price orientation

In the third phase, the MHC is now well established and thus has both a long-term perspective and is concerned with the tight integration and control of its overseas operations (Adler & Ghadar, 1989). Established MNCs frequently focus on global strategy implementation and the transfer of expatriates and organizational culture to their foreign offices (Edstrom & Lorange, 1984). However, because the firm is concerned primarily with cost efficiency of production it will still continue to value cost efficiency over long-term career development and not have an overriding interest in cross-cultural sensitivity and differentiation (Adler & Ghadar, 1989).

7.5.8 Phase 4 Price competition

In this phase the MNC faces increasing competition and change and thus to survive must develop a long-term view and a cadre of expatriates who can guide the firm with a global vision. The modern MNC must have the ability to have both an overall global strategy (integration) as well as be highly sensitive to the host country and culture (differentiation) to maintain competitive advantage (Adler & Ghadar, 1989). The MNC will need to evolve a global multicentric perspective on cultures which includes sensitivity to both national cultures as well as their subcultures in each of its overseas offices. In the next section of the paper the second stage or operational phase of the contingency model is developed to illustrate how specific IHRM practices and decisions should be made.

7.6 Operational Phase of Contingency Model

Once the critical strategic IHRM criteria have been formulated they must be carefully translated into specific IHRM decisions to ensure effective implementation. In this section the operational phase of the contingency framework will be presented for the following major aspects of IHRM policies: selection and recruitment, training and development, spouse and family considerations, performance appraisals, compensation, and expatriate career issues. Three major themes mark this section: First, U.S. IHRM practices typically lack sufficient sophistication to deal with the increasingly competitive and complex environment of the MNC. This problem is easily demonstrated by illustrating how U.S. IHRM practices lag the PLC phase of the international firm. Second, U.S. MNCs need to improve the linkages between their strategic IHRM goals and specific practices. Third, U.S. international firms can learn a great deal about how to conduct more effective IHRM operations by reviewing the experiences of European and Japanese MNCs who have more advanced overseas practices (Tung, 1988).

7.6.1 Selection and recruitment

Perhaps the most critical IHRM practice an MNC undertakes is its selection and recruitment of expatriates and/or host country nationals in the overseas operations. Tung (1981; 1988) proposes a selection process we incorporate into our contingency model on developing effective IHRM practices. Although Tung develops these factors primarily for selection, they are generally applicable to all major IHRM functions. Tung (1981; 1988) notes five aspects should be considered in the selection process: the nature of the job, the degree of cultural differences, ability and willingness of expatriates and their families to work overseas, and consideration of hiring host country nationals. To these five criteria we add a sixth-the need to consider the career plans and development of expatriate and host country nationals. It is important to recognize that the importance of these operational criteria stem from the four strategic IHRM components noted earlier - timing, cost versus development, integration, and differentiation, all of which must be considered in conjunction with each other. In turn these strategic factors influence the six operational criteria which also must be evaluated in relation to each other. We now discuss the six selection factors of the operational phase in greater detail. To enhance selection, the MNC should: (1) Identify the nature of the job to assess the degree of cultural interaction required. For example, higher level managerial positions involve a large degree of interaction with the host country and thus require effective cultural relational abilities. Additional job features which should be considered include the degree the position requires organizational experience, particularly in corporate headquarters (Cullen, 1988) and the extent to which the expatriate candidate's managerial style may match that of the host country (Spruell, 1985). Hays (1974) classifies overseas jobs into four types all of which require different levels of organizational, cultural, and work experience: executive, functional department head, trouble shooter, and rank and file member. (2) Determine the extent to which the host country's economic, political, legal, and cultural systems depart from those of the U.S., to decide how important it is to consider adaptation abilities in the selection of expatriates. (3) Assess expatriate candidates in terms of their willingness and ability to serve overseas. It is important to remember that a desire to work abroad does not always transfer to the ability to adapt to an overseas assignment (Heller, 1980). A greater number of employees are unwilling to disrupt their family, spouses, careers, and their leisure interests for international assignments (Heller, 1980) and thus their motivation to go abroad must be carefully and sensitively assessed. Important factors pertaining to the ability to adapt for overseas positions include flexibility, patience, general adaptability to different cultures (Cullen, 1988), tolerance (Spruell, 1985), and maturity (Heller, 1980). (4) Assess the ability and willingness of the expatriate candidate's spouse and family to consider living overseas. Failure of the family, particularly the spouse, to adjust to overseas assignments has been found to be the leading cause of expatriate failure (Black, 1988; Stephens & Black, 1988). (5) Consider local nationals (differentiation). Unless legal or other considerations prevail, Tung (1988) recommends that MNCs hire host or local country nationals assuming equal ability and experience. Hiring and promoting local nationals is often important to their increasing expectations and to prevent resentment of foreign (corporate) domination of overseas operations (Pucik, 1984; Tung, 1979). In addition, the lower salary cost and cultural sensitivity are other reasons firms hire local country nationals (Tung, 1979). (6) To maximize the motivation of the employee the MNC should also seek to meet career and development needs of the expatriate or local country national job candidate. In assessing local country nationals and expatriates from other nations it is also important to consider how career motivations may vary by country culture (Derr, 1986). Table 2 summarizes the expected emphasis by the MNC on these six operational criteria for each of the four PLC phases. In general, firms in Phase 1 will focus more on technical job tasks, be less sensitive to cultural differences, and develop short-term focus on the needs and skills of expatriate candidates. In contrast, by phase 4, MNCs should be highly sensitive to host country cultures, require personnel with strong corporate and cultural relational abilities, and emphasize the long-term development of expatriate and host country nationals (Adler & Ghadar, 1989; Pucik, 1984).

Unfortunately, most U.S. MNCs use neither formal selection tests nor criteria for technical, managerial competence, or cultural relational abilities (Cullen, 1988; Tung, 1988). Because U.S. firms generally lack formal selection programs, they tend to base their decisions on technical skills (Cullen, 1988), willingness to go overseas (Cullen, 1988; Tung, 1988), and short-term tax and financial considerations (Pucik, 1984). Some selection decisions to go abroad are even made because of poor U.S. performance; in essence a move on the part of the MNC to shelve or remove the particular individual (Tung, 1988), As can be observed, U.S. MNCs frequently employ early phase IHRM practices when they are engaged in a later phase (e.g. 3 or 4) of business (Adler & Ghadar, 1989).

Consistent with the contingency model, Pucik (1984) makes additional suggestions for improved staffing for firms who face greater competition and change. First, U.S. MNCs should take greater advantage of their competitive advantage of having access to a large number of foreign students in U.S. universities. Second, U.S. firms can benefit by more carefully evaluating their longer term staffing needs and seek to fill future positions in the corporation

Table 2   Continuum of Operational IHRM Criteria and Decisions by PLC Phase

Contingency Framework	Continuum of Selection and Recruitment Practices			
	Phase 1	Phase 2	Phase 3	Phase 4
1. Nature of Job	Task	Task Relational	Task MNC	Task Relational MNC
A. Task/specialty				
B. Cross-Cultural Relational Need				
C. MNC Knowledge & Experience				
2. How Different Culture is	Little	Some	Some	High
		(Degree of Consideration)		
3. Ability to Adapt Importance to MNC	Low	Medium	Medium	High
4. Spouse/Family Considerations	Low	Medium	Medium	High
5. Differentiation Consider Local Nationals	Low	Medium	Medium	High
6. Integration MNC need to Develop Cadre of Expatriates	Low	Medium	High	High

rather then always hiring for a particular position which is needed immediately. Third, greater consideration may be given in recruiting geographic area specialists and nationals rather than focusing first on technical specialties (e.g. finance, accounting, etc.) as often is the case. Realistic previews of the job have been found to be effective in recruiting in U.S. organizations (Wanous, 1980). Likewise, accurate perceptions of cultural differences of the overseas assignment (Barley, 1987) are likely to improve selection (see next section on training). Such improved programs require that personnel with sufficient international experience and training be utilized in the selection processes (Harvey, 1985).

Finally, MNCs currently have employed very few women expatriates despite the large increase in women In managerial positions and graduate business schools in the U.S. (Adler, 1984a; 1984b; Harvey, 1985), Jelinek and Adler (1988) found that many women expatriates considered their gender created more advantages than disadvantages in overseas assignments. These advantages include higher visibility, better interpersonal skills and sensitivity, higher status, and the ability to collaborate in a non-competitive situation. Furthermore, many foreign workers do not perceive U.S. women in the same restricted manner as women from their own country (Jelinek & Adler, 1988).

The state of current U.S. MNC practices indicates the lack of influence HRM managers have on senior level management and a general lack of executive management orientation towards IHRM issues (Miller, Beechler, Bhatt, & Nath, 1986). In contrast, the underlying theme of these suggestions is for the MNC to take a longer term and developmental perspective of its operations and staffing. In order to retain these employees long-term after they are selected, the MNC will need to develop effective practices in all aspects of IHRM, particularly training and development.

7.6.2 Training and development

As with selection, Tung (1988) reports that only a small proportion (32%) of U.S. MNCs have formalized training programs for expatriates, a significantly lower proportion than European and Japanese firms. Other data indicate even this U.S. figure may actually be high (Finney & Von Glinow, 1988). Intercultural training can be defined as any program which enhances an individual's ability to live and work in a foreign setting (Tung, 1981) and thus should be tailored to the specific needs of the future expatriate's job and cultural setting. Selection programs will dictate to a large degree the specific needs for training programs. Hence, the selection contingency framework can be utilized for training and development. The type, number, and intensity of training programs will be dependent on the following for both expatriate and host country nationals: (1) job requirements such as the technical aspects and cross-cultural relational need, (2) how different the culture is, (3) the willingness and ability of the expatriate to adapt different cultural situations, and (4) the need to consider longer term career and development needs for the expatriate. Based on an analysis of these factors, expatriates can obtain training in some portion of five major types of training programs identified by Tung (1981; 1988): area studies, culture assimilation, language training, sensitivity training, and field experience. Additional training needs are dictated by spouse and family considerations of the expatriate (to be discussed in greater detail in the next section). Little empirical research has been conducted on which type of training program is most effective. In general, it is believed that experiential (interpersonal) training approaches are more effective than documentary (information) methods. However, Earley (1987) found that both a detailed documentary approach and roleplaymg exercises were additive and approximately equal in effectiveness. These two approaches were found to be significantly more effective than an approach using only general information on the foreign nation. This finding provides some indication of the effectiveness of using multiple training programs and for utilizing programs which include information on specific incidents and general information rather than programs with just general information (Earley 1987).

Similar to Barley, Lee and Larwood (1983) found that satisfaction of U.S. expatriates in Korea was related to the degree they adopted Korean attitudes and cultural values. Black's (1988) research indicates that general adjustment of expatriates is related primarily to predeparture knowledge, association with local nationals, and family adjustment.

It is important to remember that cultural adjustment involves a number of dimensions such as language and communication skills, interactions, participation in host country activities, and displaying appropriate behaviors (Benson, 1978). Therefore, training programs in most instances need to address at a minimum a critical core of these aspects. Enabling the expatriate to develop a realistic preview of what the overseas employees and company (Miller, 1977) and culture (Spruell, 1985) is important to ensuring success. Such programs should also assist the expatriate in developing their own coping mechanisms for overseas living (Adler, 1986; Spruell, 1985). Because selection and training are so tightly linked a summary of the expected practices for these two IHRM functions are shown for PLC phase 1, product orientation, to phase 4, price competition, in Table 3. This continuum involves going from emphasis on the short-term and minimal selection criteria and training in phase 1 to more formal and sophisticated approaches in phase 4 which involves evaluation of long-term technical, corporate, and cultural attributes as well as consideration of the expatriate's family and local country nationals.

Thus far, the U.S. does not appear to be applying this type of contingency framework as readily as other nations. Tung (1981, 1988) reports that in general, European and Japanese programs tend to be longer, more in-depth, and more customized than those in the U.S. Furthermore, European and Japanese training programs utilize greater on-job exposure and involve emotional aspects, not just intellectual. In terms of foreign language skills there are disagreements among researchers on exactly how essential or to what extent it is necessary for the individual expatriate and the MNC overall

Table 3   Contrast of Operational Practices in Selection and Training Phase I to Phase IV Continuum

Contingency Model Consideration	Selection	Training
Phase I Product Orientation		
Nature of Job or Task Demands	Technical or Task Skills	Not required
Assess How Different Culture is	Assume Nations are Similar: Little Impact on Decision	Little impact on training
Assess Expatriate on Ability & Willingness to go Overseas	Willingness to go Overseas is Major criteria	None
Spouse and Family Considerations	Little Concern: No Formal Process	No Training or Information
Consider Local Nationals for Job	Little Concern: Consider MNC Personnel	None
Need to develop MNC Expatriates	Low Concern: Short Product focus	Minimal Consideration
Phase IV-Price Competition		
Nature of Job Demands	Relational and MNC Experience	Training as Needed for Relational Skills
Assess How Different Culture is	Assume Nations are Different	Customize Culture Training Programs
Assess Expatriate on Ability & Willingness to go Overseas	Willingness to go and Ability to Adapt	Training as Needed for Adaptation
Spouse and Family Considerations	Interview and Assess Ability and Willingness to go Overseas	Provide Training, Information and Career Assistance
Consider Local Nationals for Job	High Consideration	Evaluate Training needs
Need to develop MNC Expatriates	Consider longer term MNC and Expatriate needs	Provide Training as Needed for the person and firm

to develop extensive foreign language skills (e.g. Adler, 1986; Cullen, 1988; Heller, 1980) and this remains an important empirical question. It is clear however foreign language skills are important and that a decreased U.S. emphasis in this area can be observed: More Europeans have multilingual skills and place a greater importance on speaking a foreign language than Americans (Tung, 1988). Second, many U.S. MNCs place excessive emphasis on English as the only language in the corporation which decreases sensitivity to local country cultures (Adler & Ghadar, 1989; Pucik, 1984). A third difference lies in the approach to teaching a foreign language. The leading Japanese training institutes (e.g. Institute for International Studies and Training (IIST), Japanese American Conversation Institute (JAG), School of International Studies) believe that fluency in foreign language requires deep understanding of the foreign culture. Regarding attitudes toward different cultures there are indications that U.S. expatriates have improved their understanding of other cultures (Cullen, 1988), but are still often viewed as being less tolerant of foreign practices than nations with greater overseas experiences such as Britain (Tung, 1988). Americans tend to have neither the multi-national experience, historical perspective of different cultures, nor the extensive formal training of some cultures, and thus are often ill-prepared for overseas assignments. Furthermore, only about one-third of MNCs that do have formal training programs actually evaluate the effectiveness of them (Tung, 1988). Clearly critical self-appraisal and feedback is needed before U.S. MNCs can develop effective international training programs.

In addition to training expatriates on local country languages and cultures it is critical for MNCs to develop the corporate orientation and long-term strategic and global orientation of its managers and professional staff (Adler & Ghadar, 1989; Lee & Larwood, 1983; Pucik, 1984). One of the current problems of U.S. international firms is that they develop too few expatriates with broad global perspectives and fail to orient the development of expatriates specifically to the strategic objectives of the MNC (Pucik, 1984). Similarly, another need is for U.S. MNCs to implement global information systems on market and organizational factors. A worldwide development emphasis is needed to create a common global perspective on the organization's strategy (Pucik, 1984).

U.S. MNCs tend to neglect the training of not only expatriates, but also local country nationals (pucik, 1984). It is critical for international firms to develop their local country nationals to provide for effective integration with the corporate overseas headquarters. It is critical for training and development programs to be culturally sensitive or else transfer skills may not be effectively developed or desired by the trainees (Putti &, Yoshlkawa, 1984). Furthermore, in terms of long-term development, most executives and senior managers of MNCs continue to come from the corporate headquarters in the U.S. (Ondrack, 1985; Pucik, 1984) as well as Japan (e.g. Tung, 1982). This ethnocentric perspective is likely to hamper the motivation of host country managers and limit the global vision of the MNC (Ondrack, 1985).

Tung's (1988) survey shows that U.S. firms with specific selection criteria and formalized training programs have significantly lower expatriate failure rates than firms without these practices. Failure rates occur when poor performance or personal reasons force the company to recall expatriates early or fire them from their overseas assignment. One suggestion for improving training is for corporations to send some younger employees when the compensation requirements and needs are much lower than those of older, more established expatriates (Pucik, 1984). Another possibility is to emulate the approach taken by the U.S. Peace Corps. The Peace Corps has developed a comprehensive four-step training process which involves a sophisticated assessment method for assigning candidates, preservice training of 10-14 weeks in the host country, in-service training while the individual is abroad, and a close-of-service training workshop prior to returning to the U.S. (Barnes, 1985). Another interesting alternative in training is for MNCs to develop arrangements where employees are trained overseas in other independent foreign companies (Pazy & Zeira, 1983; Zeira & Pazy, 1985). Zeira & Pazy (1985) studied 17 MNCs who used this joint venture technique and found that this method of on-the-job, in-house training is particularly effective in developing crosscultural awareness and sensitivity and in decreasing their own cultural biases. Furthermore, organizational change and development can occur when a group of employees are sent to an overseas company for training. While this approach may be useful only to certain MNCs it certainly possesses a number of advantages over more traditional training programs such as developing long-term relationships between companies, on-the-job training, and potential for organizational development and learning.

In considering training programs it is important to remember that more formal selection and training practices are costly (Pucik, 1984) and MNCs need to carefully evaluate the short-term costs versus long-term benefits. Thus far most U.S. MNCs have focused more on the costs of expatriate training programs which are known and not on the benefits which are difficult to quantify. Again, the focus is more on the symptoms of the problem than on the source of the problem (Harvey, 1983). However, in the future the extremely high cost of expatriates (Cullen, 1988; Pucik, 1987), estimated high failure rate of expatriates of 30% (Henry, 1965) and 10-20% (Tung, 1982), and the reduced effectiveness of expatriates who stay on, should provide greater incentive for U.S. firms in the future to conduct more systematic and training procedures. Another facet of training is how well the expatriate's spouse and family adjust to the overseas assignment which will be discussed next.

7.6.3 Spouse and family considerations

One of the central reasons for expatriate failure is due the inability of the expatriated family and spouse to adjust to the overseas setting (Adler, 1986; Hays, 1974; Stephens & Black, 1988; Tung, 1988). In most cases overseas placements are more difficult for the spouse than for children who are younger and more adaptable. Similarly, it is often easier for the expatriate to adapt since he or she often is in a work setting which involves the use of some English and is shielded to some degree from the surrounding environment. Spouses often face greater difficulties and culture shock because they have to deal more extensively with the foreign culture and do not have extensive contact and friendships which are developed by the working spouse with English speaking people at work (Adler, 1986; Harvey, 1985). Frequently overseas assignments are difficult in the early stages, but the excitement of being in a foreign country often produces initial positive feelings. After the initial transition, the more difficult task of facing the realities of overseas living emerges.

Later more positive feelings develop once the expatriate and his or her family more fully adjust (Adler, 1986; Harvey, 1985; Stephens & Black, 1988). The need for specific selection and training programs for the spouse and the family is dependent on the following factors (1) how different the culture is, (2) the ability and willingness of the spouse and family to live overseas, and (3) the perceived need of the company to invest in the longer term development of the expatriate. Unfortunately, few U.S. MNCs have formal policies on training spouses or families for overseas placement (Finney & Von Glinow, 1988; Stephens & Black, 1988; Tung, 1988). Most companies do not offer training programs and those that do generally provide only language training classes, not cultural adjustment programs (Adler, 1986). However, about 50% of U.S. firms now recognize that interviews with spouses of potential managerial expatriates need to be included in the selection process (Stephens & Black, 1988). With the increasing number of women in the workforce another concern is how to accommodate dual career needs of expatriates. Most U.S. companies do not have formal policies to assist spousal employment abroad. This is true even though research indicates dual career couples adjust as well overseas as single career couples (Stephens & Black, 1988) and that spousal discontent can affect the repatriation process (Harvey, 1989). It is clear that more formalized programs for selection, training, and coordination of spouses and families overseas is warranted for more effective expatriate adjustment and may be the most effective way for MNCs to decrease expatriate failure (Hays, 1974). The single most important goal of such programs should be to create a meaningful life for spouses during their stay abroad (Adler, 1986). For example, some European companies have formal processes where returning expatriate families discuss their experiences and give advice to outgoing expatriate families (Tung, 1988). Helping spouses to continue developing their careers while overseas is also receiving new emphasis (Adler, 1986). Thus, it is critical to assist the adjustment of the spouse and family before arriving in the foreign nation (Harvey, 1985) as well as upon return to their home country (Harvey, 1989). In addition to developing effective selection and training programs for expatriates and their spouses it is also important to implement performance appraisal systems which allow for adequate adjustment to the overseas assignment.

7.6.4 Performance appraisal

As in selection and training programs for expatriates and their families, the MNC's performance appraisal system needs to be flexible and adaptable. This flexibility can be developed by incorporating the operational contingency advocated previously. First, the nature of the expatriate's job tasks and must be carefully considered. Expatriates often have much different greater work responsibilities and autonomy abroad than they do in the U.S. In addition, many expatriates go from more technical positions in the U.S. to overseas positions which require greater strategic management, diplomatic, and inter-personal skills (Phatak, 1983; Pucik, 1984). Second, the extent of differences in culture requires varying degrees of adjustment. Third, people possess different timetables for adjusting to overseas assignments; some adjust more quickly than others.

The combination of dealing in a new cultural setting and having expanded work responsibilities creates a situation where expatriates frequently do not perform effectively for the first year or even longer. Generally, corporate headquarters are not sensitive to the special problems faced by expatriates (Spruell, 1985) and expect performance similar to what would be in the U.S. (Harvey, 1983). Thus, performance appraisals should often be conducted well after the first year of duty (Pucik, 1984; Tung, 1988), carefully reflect the strategic nature of the position (Pucik, 1984), and adjusted periodically over the expatriate's tenure overseas.

The difficulty in accurately assessing the contribution of the expatriate managers arises in how to define and measure the performance. Pucik (1984) observes it is frequently difficult to measure the financial performance of an overseas subsidiary due to difficulties in exchange rate fluctuations, special financial considerations involved with new subsidiaries, long distances, and infrequency of contact between overseas offices and headquarters. Ultimately appraisal systems must be tailored to the special needs of expatriates and their overseas situations and involve a careful blending of short-term and long-term goals (Pucik, 1984).

An additional set of considerations involves the integration and differentiation needs of the MNC. Appraisals need to be similar enough across the MNC for equity and comparison purposes. On the other hand, appraisals should be sensitive to cross-cultural differences. For example, U.S. appraisals tend to formal and explicit, while in Japan performance evaluations are informal and implicit (Pucik, 1987; Von Glinow & Chung, 1989). Many other nations lack sophisticated appraisals altogether. Thus, appraisals must be must reflect the specific needs of the expatriate, organization, and host country culture.

7.6.5 Compensation and reward systems

Appraisal systems play an important role in determining compensation in U.S. international firms and should ensure that performance is linked to corporate strategy so that employees receive appropriate levels of rewards (Pucik, 1984). Complicated and intricate financial compensation packages for expatriates have been developed by many MNCs (e.g. Phatak, 1983). These formulas are beyond the scope of this paper. For this reason we will not make an extensive discussion of compensation at this time except for how the different problems and features relate to the contingency model. Four factors which play a major role in determining compensation schemes include the nature of the expatriate's job responsibilities, how different the host country culture is, spouse and family considerations, and salary levels of local nationals. A critical challenge is to develop a compensation system which is both equitable to local nationals, expatriates, and parent country employees, and has sufficient incentives to attract, motivate, and retain candidates to overseas settings (Phatak, 1983; Pucik, 1984). In general, a centralized compensation administration is more effective in developing an equitable and consistent system, while a decentralized program is more likely to be sensitive to local conditions (Toyne & Kuhne, 1983). Toyne and Kuhne (1983) found that most U.S. MNCs use a centralized compensation administration structure for policies, procedures and control. However, the compensation administration is generally geocentric or sensitive to the overseas offices.

Furthermore, certain aspects of compensation which are more sensitive to local environments, such as financing, tended to be delegated more to the overseas subsidiary. Overall, there is a trend toward greater centralization of compensation programs in the corporate headquarters (Toyne & Kuhne, 1983), a finding which is suggestive of MNCs moving towards at least PLC phase 3. In the long term (e.g. PLC phases 3 and 4) a sophisticated global compensation system will probably be most effective (Phatak, 1983). The global compensation system involves having the same pay for the same job in all countries regardless of whether the position is staffed by an expatriate from the nation of the MNC, local country national, or expatriate from another country (third country expatriates). Compensation differentials are then added to the base pay based on the country. However, most MNCs are not currently at this level of sophistication (Phatak, 1983). As with performance appraisal, a crucial issue in compensation is how to reward performance when expatriates make strategic contributions to the subsidiary and MNC. It is particularly difficult to achieve this linkage when MNC appraisal systems must deal with such factors as cultural differences, difficulties in determining the strategic and financial performance of the overseas units, and the large physical distances to the corporate or divisional headquarters (Pucik, 1984). Another problem is that most MNCs do not offer effective nonfmancial reward programs (Pucik, 1984). For example, most MNCs still rely on promotions as the critical reward, but these are often only provided to parent country expatriates (Pucik, 1984) and ignore the fact that many employees desire different types of rewards other than upward movement (Adler, 1986; Derr, 1986). Thus, when positions for upward movement are limited it may be useful for MNCs to offer challenging lateral moves for expatriates to enhance their motivation and long-term development (Pucik, 1984). Developing culturally sensitive reward systems is particularly important for the employment of local country nationals and expatriates from other countries. Hofstede (1980, 1984) has found that motivation varies significantly based on two of the four work dimensions: masculinity versus femininity and low versus high uncertainty avoidance. For example, employees in Sweden tend to score high on the femininity side and thus desire non-financial rewards such as autonomy, interpersonal relations, and time off. In contrast, nations like the U.S. which score higher on the masculinity scale, place a higher value on financial incentives, status, and job challenge. Thus, reward systems must be tailored carefully to meet equity, development, and cross-cultural needs within the MNC. These various needs will be discussed in further detail in the following section on meeting the career needs of expatriates and host country nationals.

7.6.6 Career and repatriation needs

Career and repatriation considerations are closely linked to those of reward systems and all of the other IHRM programs. Effective IHRM policies need to be developed for the entire period of overseas duty as well as repatriation. In selection and recruitment, both assessing and meeting career needs can assist the individual employee and organization in achieving a better person/organization fit in U.S. firms (Schneider & Schmidt, 1987) as well as international firms (ICanungo & Wright, 1983). As noted earlier expatriates often have much greater work responsibilities and autonomy abroad than they do in their home countries resulting in adjustment difficulty. One way to facilitate work adjustment is to provide a comprehensive network of overseas contacts with incoming and outgoing expatriates (Black, 1988; Tung, 1988). Because of the great changes in overseas assignments it is important that expatriates find there own unique way or receive training to manage stress effectively for both work and personal living situations (Adler, 1986). In terms of development and reward systems, one of the most fundamental issues facing U.S. companies is how to integrate the personal and career needs of expatriates with the longer term goals of the MNC (Pucik, 1984). As mentioned earlier, promotions for local country nationals and third country expatriates still lag substantially behind those of home country expatriates. As a result, the MNC gradually loses the most effective local personnel over time, or those who stay have declining organizational commitment. Partly as a result of this, MNCs are occasionally required to quickly hire ineffective local employees at an increased wage rate (Pucik, 1984). A related issue is that long-term expatriate stays are still often viewed as harmful to career mobility in U.S. organizations (Cullen, 1988; Phatak, 1983; Stephens & Black, 1988). This is known as the "out-of-sight, out-of-mind" syndrome (Tung, 1987). Expatriate experiences are still often not valued highly by U.S. firms; overseas assignments place the individual outside the traditional promotion route and contacts at the company headquarters. Oddou, Mendenhall, and Bedford (1988) studied the careers of over 100 expatriates and found that expatriation had a very positive impact on their careers, while repatriation often had a neutral or negative effect. Only about one-half of returning expatriates are promoted upon return (Adler, 1986). Frequently, expatriates are placed in mediocre positions upon return to the parent country and typically experience difficulties such as working under tighter organizational constraints of the parent country, loss of status, excitement, and authority, new financial burdens, and reentry adjustment (Adler, 1986; Harvey, 1989). These factors can create executive stress and psychological tension (Harvey & Lusch, 1982), but many MNCs are not aware of these problems and issues (Tung, 1987). Typical reentry problems occur because the employee, country, and company have all changed and some describe repatriation as being more difficult than going overseas. Frequently, it is difficult for the expatriate to apply his or her knowledge and experience from overseas because other employees in the parent company have not had a cross-cultural experience (Adler, 1986). Often the employee finds that his or her technical skills are now obsolete (Tung, 1987). Often spouses have a difficult time readjusting as well upon return. For these reasons approximately 20 percent of expatriates desire to leave their company upon repatriation (Adler, 1986) which is a significant loss of investment and experience for the firm. Unfortunately, poor experiences with repatriation are often magnified across the company, creating even greater concerns and anxieties about international assignments (Tung, 1984; 1988). Harvey (1986) surveyed personnel administrators on their repatriation policies and found that only 31% of the MNCs had formal programs. The major reasons cited for the lack of such activities were the lack of expertise in establishing the program, cost of program, and no perceived need for such training by top management. Many of the personnel administrators were not certain how to develop a repatriation program and who should be involved in it. It also appears that MNCs whose operations require less involvement and sensitivity to foreign cultures (e.g. industrial products companies) were less likely to have formal policies (Harvey, 1986). These problems again poignantly illustrate how MNCs are still at PLC phase 1 or 2 thinking, despite the fact that they are facing phase 3 or 4 problems and situations. One recommendation is for MNCs to consider adopting some of the policies of European companies such as charging Personnel Department professionals with overseeing the career concerns of the expatriates or develop a mentoring system where the expatriate is paired with a superior at corporate headquarters (Phatak, 1983; Spruell, 1985; Tung, 1988). Other practices for facilitating repatriation include written agreements to limit the number of years spent abroad and promising promotions upon return, taking additional measures to keep the expatriate in contact with corporate headquarters by scheduling regular meetings with travel to corporate headquarters, and to assist in paying real estate and legal fees to decrease the financial burden (Phatak, 1983).

Adler (1986) recommends that companies provide debriefing sessions, reentry training programs, identification of specific skills learned abroad, and recognition of the expatriates work overseas and future value to the company to facilitate repatriation, Spruell (1985) notes that sending only top performing employees abroad helps ensure that expatriates will be promoted and valued upon repatriation. Tung (1987) recommends that MNCs develop an integrated career planning program which functions from prior to the overseas assignment through to repatriation. Recommendations by MNC personnel administrators on developing ideal repatriation programs include financial and tax assistance and career path assistance (Harvey, 1986). It is important to note that these MNCs placed less emphasis on the psychological aspects of reentry despite their importance.

Ultimately, the key is for MNCs to place a greater emphasis on the value of overseas assignments for the organization and the individual. For the organization it is important that overseas experience be viewed as important to its long-term and strategic needs. For the expatriate the overseas assignment should be valued in terms for both the development and career path of the individual and not seen as just a mechanism to serve the immediate job needs of the company (Adler, 1986). At a time when the MNC's competitive advantage may well be its human assets, it is noted that expatriate assignments often allow greater challenge, responsibilities, and decision-making opportunities (Cullen, 1988; Stephens & Black, 1988) as well as potential to affect the company (Tung, 1988) than positions in the U.S. The inconsistency in the U.S. MNC's treatment of expatriation must be addressed. If expatriate assignments are viewed as attractive (as they are in Japan and Europe) productivity increase and repatriation problems lessen. The long-term employment orientation in many European and Asian companies assists in making overseas positions highly valued because employees are more willing to travel for the company when they see overseas assignment as instrumental to promotion. This spirit of internationalism also benefits the MNC by creating a greater dedication and loyalty of employees to company objectives (Tung, 1988). In addition, an experienced group of expatriates can help the company obtain a global vision and orientation which will be increasingly important in the future (Pucik, 1984). Because of the high costs associated with expatriation it is clear that companies will gain in both the short and long-term by carefully developing formal, systematic programs for the selection, training, and ongoing career needs of expatriates. There is a great deal of common ground for such activities to meet the needs of both employees and the organization.

7.7 Implications for MNC Success

As U.S. MNCs face an increasingly competitive and changing environment, they need to more quickly adapt many critical aspects of the company, particularly their IHRM practices. A two-step contingency model is proposed showing how the MNC can develop effective strategic and operational IHRM practices. A detailed illustration of this model is shown in Figure 2. In the strategic phase,

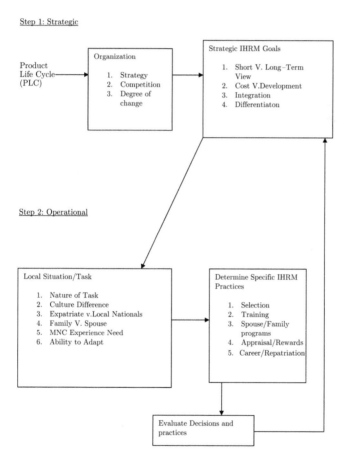

Figure 2 Two-Step Contingency Model for Developing Effective IHRM Practices

the international PLC is shown to influence the major environmental and organizational factors of the MNC which in turn affects the firm's strategy. The role of senior management and the Personnel department is then to translate the MNC business strategy to strategic IHRM objectives (Miller, Beechler, Bhatt, & Nath, 1986; Pucik, 1984). Four fundamental strategic IHRM objectives involve short-term versus long-term planning, cost versus development, and the need for integration and differentiation. The operational phase of the contingency model involves translating the strategic IHRM objectives into specific decisions. Based on the work of Tung (1988) a formal and systematic contingency framework is developed at the operational level using the following six criteria: nature of job or task, how different the host country's culture is, the ability of the expatriate to adapt, spouse and family considerations, consideration of the use of host country nationals, and the need for longer term development of expatriates. As shown in Figure 2 it is also important for the MNC to continually assess the effectiveness of these various IHRM programs and adjust them to changing conditions accordingly.

Not only is it is essential for firms to develop IHRM strategies appropriate to its specific PLC phase, it is also important to realize that the rapidly changing and competitive global economy is pushing many U.S. firms quickly into advanced PLC stages where high degrees of both integration and differentiation are required (Adler & Ghadar, 1989; Galbraith, in press; Schoonhoven & Jelinek, in press; Von Glinow & Mohrman, in press). One clear implication is that U.S. IHRM practices need to be more formally and systematically developed. Accordingly, a number of suggestions for the improvement of IHRM practices at the operational level were also offered in this manuscript. These suggestions which pertain to selection, training, spouse and family, performance appraisal, reward systems, expatriate career, and repatriation aspects are summarized in Table 4.

Getting U.S. firms to employ the above advocated contingency framework and to begin addressing deficiencies in their IHRM practices is a difficult proposition. A shift in this direction will probably not occur until three events transpire. First, senior u.s. managers must experience sufficient discontent with the results of the current system. Second a sufficient number of employees must possess certain minimal levels of international experience and abilities (Phatak 1983). Third, the firm naturally evolves a greater cross-cultural customers, operations, and products which in turn demand a more culturally diverse orientation (e.g. a shift to PLC phase 3 or 4) (Adler & Ghadar, 1989). This may precipitate a basic paradigm shift on how they manage human resources in general and how they fundamentally view the role and importance of

Table 4 Summary of Suggestions for Effective IHRM Practices

Area	Potential Solutions
Selection and Recruitment	1. Develop Formal criteria
	2. Implement Systematic Process
	3. Assess Willingness and Ability to go abroad
	4. Assess Task and Cultural Relational skills
	5. Recruit foreign students from U.S. universities
	6. Recruit for general geographic area expertise
	7. Realistic Job preview of overseas positions
	8. Consider local country nationals
	9. Realistic job previews of overseas assignment
	10. Consider more U.S. women
Training and Development	1. Consider Five General Types of Training Programs
	2. Develop In-depth and Tailored Programs to expatriate and country
	3. Utilize multiple training programs
	4. Assist Expatriate in Developing Coping Mechanisms
	5. Create Realistic Preview of Overseas Assignment
	6. Development Programs for Expatriate's Career Needs and for long-term
	needs of the MNC
	7. Training Throughout Overseas Stay, Including Repatriation
	8. For Cost Reasons Consider Training Younger Employees
	9. Consider Training Employees at Independent Foreign Companies
	10. Include Local Country Nationals in Training Programs
	11. Evaluate Programs
Spouse and Family	1. Interview for Future Adjustment
	2. Consider Dual Career Needs
	3. Cultural Training Programs
	4. Interact with other Expatriate Families
	5. Prepare Spouse/Family for reentry to US.
Performance Appraisal	1. Delay Performance Appraisal
	2. Consider local country cultures in Developing Appraisals
	3. Integrate and Carefully Measure short term and Long-term Performance
	criteria and Goals
Compensation & Reward Systems	1. Develop Global System of Compensation by Job Titles
	2. Consider local country cultures in Developing Rewards
	3. Develop non-financial rewards
	4. Balance Equity Needs within the MNC and Sensitivity to Local Country Cultures
Career & Repatriation	1. Network of Overseas Contacts
	2. Longer-term Stays and Specify Tenure of Assignment
	3. Promotions Sensitive to Career Needs of Both Expatriates and Local Country Nationals
	4. Mentoring Program to Assist Career Development
	5. Corporate Emphasis on Overseas Experience: Integrate Career Plans

Table 4 Continued

Area	Potential Solutions
	6. Personnel Department to Assist in Repatriation 7. Consider Career Needs and Overseas Experience of the Expatriate upon Return to the U.S. 8. Assist Expatriate and Family with Job, Financial, and Psychological Needs Upon Repatriation

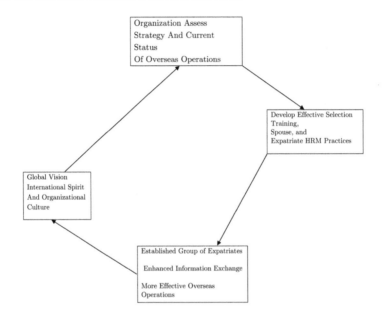

Figure 3 Evolution of Organization IHRM Practices and Global Vision of the Firm

their international operations. Figure 3 illustrates how this shift can come about.

As shown in Figure 3 as the MNC senior managers experience fundamental problems they begin to place greater emphasis on IHRM and the development of business from an international perspective. As this perspective begins to emerge, management will create more long-term oriented and effective IHRM practices. As the firm develops a more global perspective it will retain a group of expatriate managers who will return to corporate offices or staff positions in other countries. Sending only higher performers overseas also sends a message throughout the company and increases interest in overseas work (Spruell, 1985). In turn these expatriates can facilitate international thinking, contributing to an organizational culture which values international assignments. A cadre of experienced global managers can enable the MNC to respond to global threats and opportunities and create a truly global long-term business strategy (Ondrack, 1985; Pucik, 1984). Organizations and management teams with greater cultural diversity can be more creative and effective than those from a homogenous culture if managed correctly (Adler, 1986). The creation of a global vision for the firm requires the development of more effective IHRM practices to compete in an increasingly competitive world economy. The changes discussed are only in their infancy in most U.S. companies at this point. The above suggestions toward enhancing MNC success through IHRM practices can also serve as a research agenda for scholars and practitioners alike. Very little empirical or theoretical work currently exists on modeling IHRM practices along PLC phases. We believe that effort must be expended in this area, if research is to keep abreast of practice in this complicated international milieu.

References

1 Adler, N. (1984a) Women in international management. California Management Review, XXVI. Summer, 78–89.
2 Adler, N. (1984b) Expecting international success: Female managers overseas. Columbia Journal of World Business, Fall 77–83.
3 Adler, N. (1985) International Dimensions of Organizational Behavior. Boston, MA: Kent Publishing.
4 Adler, N. and F. Ghadar. (1989) Globalization and human resource management. In Research in Global Strategic Management: A Canadian Perspective. Alan Rugman (Ed) Volume 1 Greenwich, Conn: JAI Press.
5 Barnes, R.L. (1985) Across cultures: The Peace Corps training model. Training and Development Journal, October, 46–49.
6 Benson, P.G. (1978) Measuring cross-cultural adjustment: The problem of criteria. International Journal of Intercultural Relations, Spring, 21–37.
7 Black, J.S. (1988) Work role transitions: A study of American expatriate managers in Japan. Journal of International Business Studies, .19. Summer: 277–294.
8 Conway, MA. (1984) Reducing expatriate failure rates. Personnel Administrator. 29, July, 31–32, 37–38.
9 Cullen, T.P. (1988) A profile of the American expatriate manager in Japan. Paper presented at the Annual Academy of Management Conference. August, 1988, Anaheim, CA.

10 Davis, J., Ken; S. & Von Glinow, MA. (1987) is the Japanese management craze over? International Journal of Management. 4, September, 486–95.

11 Derr, C. B, (1986) Managing the New Careerists. San Francisco, CA: Jossey-Bass Publishers.

12 Doz, Y. & Prahalad, C.K. (1981) An approach to strategic control in MNCs. Sloan Management Review. Summer, 5–13, and Fall 1981, 15–29.

13 Earley, P.C, (1987) Intercultural training for managers: A comparison of documentary and interpersonal methods. Academy of Management Journal, 30, 685–698.

14 Edstrom, A. & Galbraith, J, (1977) Transfer of managers as a, coordination and control strategy in multinational firms. Administrative Science Quarterly. 22, 248–263.

15 Edstrom, A. & Lorange, P. (1984) Matching strategy and human resources in multinational corporations. Journal of International Business Studies. 15, Fall 125–137.

16 Evans, P.A.L. (1987) Strategies for human resource management in complex MNCs: A European perspective. In V. Pucik's Academy of Management Proposal "Emergent Human Resource Strategies in Multinational Firms: A Tricontinental Perspective," 9–11.

17 Finniey, M. & Von Glinow, M.A. (1988) Integrating academic and organizational approaches to developing the international manager." Journal of Management Development. 7: 16–27.

18 Galbraith, J. (in press) Technology and global strategies and organizations. In M.A. Von Glinow & S. Mohrman (Eds) Managing Complexity in High Technology Organizations, N.Y.: Oxford University Press, 37–55.

19 Harvey, M.C. (1981) The other side of foreign assignments: Dealing with the repatriation dilemma. Columbia Journal of World Business, 17, Spring, 53–59.

20 Harvey, M.G. (1983) The multinational corporation's expatriate problem: An application of murphy's law. Business Horizons. January-February, 72–78.

21 Harvey, M.G, (1985) The executive family: An overlooked variable in international assignments. Columbia Journal of World Business. Spring, 84–91.

22 Harvey, M.G. (1989) Repatriation of corporate executives: An empirical study. Journal of International Business Studies. Spring, 131–144.

23 Harvey, M. & Lusch, R. (1982) Executive stress associated with expatriation and repatriation. Academy of International Business Proceedings, December, 540–543.

24 Hays, R.D. (1974) Expatriate selection: Insuring success and avoiding failure. Journal of International Business Studies, 5, Spring, 25–37.

25 Heller, J.E. (1980) Criteria for selecting an international manager. Personnel. May-June, 47–55.

26 Henry, E.R. (1965) What business can learn from Peace Corps selection and training. Personnel, 41, July-August 1965.

27 Hofstede, G, (1980) Culture's Consequences: International Differences in Work-Related Values. Beverly Hills, CA: Sage Publications.

28 Hofstede, G. (1984) The cultural relatively of the quality of life concept. Academy of Management Review. 9, 3: 389–398.

29 Jaeger, A.M. 1982. Contrasting control modes in the multinational corporation, International. Studies of Management and Organization. Spring 1982, 59–82.

30 Jelinek, M. & Adler, N. (1988) Women: World-class managers for global competition, Academy of Management Executive, H 11–19.

31 Kanungo, R.N, & Wright, R.W. (1983) A cross-cultural comparative study of managerial job attitudes. Journal of International Business Studies, XIV, Fall 115–130.

32 Lee, Y. & Larwood, L. (1983) The socialization of expatriate managers in multinational firms. Academy of Management Journal 26, 657–665.

33 Lorange, P. (1986) Human resource management in multinational cooperative ventures. Human Resource Management, 25, 91–102.

34 Mendenhall, M.E» & Oddou, G.R. (1985) The dimensions of expatriate acculturation: A review. Academy of Management Journal 19, 39–47.

35 Miller, E.L. (1977) Managerial qualifications of personnel occupying overseas management positions as perceived by American expatriate managers. Journal of International Business Studies, Spring/Summer, 57–69.

36 Miller, E.L., 5. Beechler, B. Bhatt, and R. Nath. (1986) The relationship between the global strategic planning process and the human resource management function. Human Resource Planning. 9: 9–23.

37 Oddou, G., Mendenhali, M. & Bedford, P. (1988) The role of an international assignment on an executive's career: A career stages perspective. Paper presented at the Academy of International Business Conference, San Diego, CA., October, 1988.

38 Ondrack, D. (1985) International transfers of managers in North American and European MNEs. Journal of International Business Studies. Fall 1–19.

39 Pazy, A. & Zeira, Y, (1983) Training parent-country professionals in host-country organizations. Academy of Management Review. 8, 262–272.

40 Peters, TJ. and R.H. Waterman, R.H. f 1982) In Search of Excellence; Lessons From American's Best-Run Companies. New York: Harper and Row.

41 Phatak, A.V. (1983) International Dimensions of Management. Boston, MA: Kent Publishing Company.

42 Pucik, V, (1984) The international management of human-resources. In C.J. Fombrun, NJVL Tichy, & M.A. DeVahna (Eds) Strategic Human Resource Management. New York: John Wiley & Sons.

43 Pucik, V. (1987) Joint ventures with the Japanese: The key role of HRM. Euro-Asia Business Review, 6, October, 36–39.

44 Putti, J. & Yoshikawa, A. (1984) Transferability - Japanese training and development practices. Proc. Adac. Int.Bus,.Mtg., Singapore, 300–10.

45 Riggs, H.E. (1983) Managing High-Technology Companies. Belmont, CA: Lifetime Learning Publications.

46 Rugman, A.3VI. (1988) Multinational enterprises and strategies for international competitiveness. In Advances in International Comparative Management. 3. Greenwich, Connecticut; JAI Press, 47–58.

47 Schneider, B & Schmidt, N. (1987) Staffing Organizations. (2nd ed), Glenview, Illinois: Scott, Foresman and Company.

48 Schoonhoven, K. & Jelinek, M. (in press) Dynamic tension in innovating high-technology firms: Managing rapid technological change through organization structure. In M.A. Von Glinow & S. Mohrman (Eds) Managing Complexity, in High Technology Organizations. N.Y.: Oxford University Press.

49 Spruell, G. (1985) How do you ensure success of managers going abroad? Training and Development Journal December, 22–24.

50 Stephens, O.K. and S. Black. (1988) International transfers and dual-earner couples: The influence of the spouse. Paper presented at the Annual Academy of Management Meeting, August, 1988, Anaheim, CA.

51 Toyne, B. & Kuhne, RJ. (1983) The management of the international executive compensation and benefits process, Journal of International Business Studies, Winter, 37–50.

52 Tung, R. (1979) U.S. multinationals: A study of their selection and training procedures for overseas assignments. Academy of Management Proceedings, 298–301.

53 Tung, R. (1981) Selection and training of personnel for overseas assignments. Columbia Journal of World Business, 16, Spring, 68–78.

54 Tung, R, (1982) "Selection and training procedures of U.S., European, and Japanese multinationals." California Management Review. 2.5:1: 57–71.

55 Tung, R, (1984) Strategic management of resources in the multinational enterprise. Human Resource Management. 23,129–144.

56 Tung, R. (1987) Career issues in international assignments. Academy of Management Executive. I], 241–244.

57 Tung, R. (1988) The New Expatriates: Managing Human Resources Abroad. Cambridge, Mass: Harper & Row, Publishers, Inc.

58 Vernon, R. (1966) International investment and international trade in the product cycle. Quarterly Journal of Economics, 80. 190–207.

59 Von Glinow, M.A. (1988) The New Professionals: Managing Today's High Tech Employees. Cambridge, MA: Ballinger.

60 Von Glinow, M.A. & Teagarden, M. (1988) The transfer of human resource management technology in Sino-U.S. cooperate ventures: Problems and solutions. Human Resource Management, 27, Summer, 201–229.

61 Von Glinow, M.A. & Chung, B.J. (1989) Comparative human resource management practices in the U.S., Japan, Korea, and the People's Republic of China. In K. Rowland (Ed) JAI Supplement on International Human Resources Management. Greenwich, Conn: JAI Press.

62 Von Glinow, M.A. & Mohrman, S. (In press) Managing Complexity in High Technology Organizations, N.Y.: Oxford University Press.

63 Wanous, J, (1980) Organizational Entry. Reading, MA: Addison-Wesley Publishing Co.

64 Zeira, Y. & Pazy, A, (1985) Crossing national borders to get trained. Training and Development Journal. October, 53–57.

8

Development of a Knowledge Management Framework within the Systems Context

Roberto Biloslavo[1] and Max Zornada[2]

[1]*Faculty of Management Koper, Slovenia*
[2]*Adelaide Graduate School of Business,*
University of Adelaide, Australia

8.1 Abstract

Effectiveness of knowledge management depends on how knowledge management processes are aligned with an organisation's infrastructure and processes, in a manner that supports the achievement of an organisation's goals. To understand and represent theses relationships a simple list of elements and processes is inadequate. We need a holistic framework where all are integrated into a dynamic coherent whole. The proposed framework is particularly focused on dividing the identified organisational building blocks into their constituent elements along both time and content dimensions so as to define characteristics that these elements, and the relationships between them need to have to form a social ecology in which people effectively create, share and use knowledge. In this way the developed framework can assist management to understand the true nature of the relationships that exist between an organisation and knowledge management processes, and to exploit them for an organisation's success.

Keywords: Knowledge management, Knowledge management frameworks, Strategic management, Systems thinking.

137

8.2 Introduction

After a decade of intense interest, many different frameworks in the field of knowledge management have been developed by academics, consultants and practitioners. For example, Heisig (2002) described thirty different knowledge management frameworks, Rubenstein-Montano et al. (2001) twenty-six, and Holsapple and Joshi (1999) five broad frameworks. Yet at the same time, we don't have a generally accepted framework and existing frameworks need, at the very least, some refinement (Rubenstein-Montano et al, 2001; Holsapple and Joshi 1999). For this reason, some authors recommend approaching knowledge management using systems thinking (Rubenstein-Montano et al. 2001). An evolved version of the systems thinking approach can be already found in strategic management, particularly in the so-called European management school. Probably the most famous representative of this school is the St. Gallen integral management model (Bleicher, 1995; Gomez and Zimmermann, 1993; Schwaninger, 1994; 2001) but similar ideas can be found also in the model developed by Tavcar (1999; 2003). After a detailed analysis of the two mentioned strategic management models we identified three strategic pillars of an organisation and divided them along three time dimension. Identified strategic pillars are: assets, businesses, and orderliness. These were the elements that were linked to four knowledge management processes: creation, storing, transfer, and application of knowledge. As systems theory suggests is not with a detached analysis of some particular elements that we can understand the true behaviour of a system, but with a comprehensive synthesis of them. As knowledge management processes and strategic pillars per se are not sufficient for an organisation's success, we need to align and integrate them with an organisation strategy. On the other hand as the knowledge management literature suggests, an organisation needs to be focused on the right combination of people and technology to achieve the best results from its efforts. For this reason we added to the develop framework people and technology as the main building blocks of knowledge management, and two general knowledge strategies as exploitation and exploration. In this way we have gone beyond a simple list of elements to a representation of an organisation as a purposeful knowledge enabling system, In the remainder of this paper we will first delineate the main conceptual ideas behind the proposed framework. On the basis of presented findings a systems framework of organisational elements and knowledge management processes will be developed including all relevant relationships that exist between them. In the last section we will present ideas for future research.

8.3 Knowledge in an Organisation

That knowledge is of fundamental importance for organisations of any size and industry is no longer a question (Martin 2000, 17). Even if knowledge is not the sole element for an organisation's survival, it is the most important one because it supports all others (Rastogi 2002). For this reason, it is not surprising that business and academic communities are very deeply involved in understanding knowledge, and developing knowledge management processes and systems to exploit opportunities that knowledge offers to organisations. In spite of the increasing research of knowledge and related subjects, no unified definition of "knowledge" can be found in business and academic literature. Some definitions of knowledge that can be found in the literature are:

- Knowledge is a fluid mix of framed experience, values, contextual information, and expert insight that provides a framework for evaluating and incorporating new experiences and information. (Davenport and Prasak 1998, 5).
- Knowledge is the power to act and to make value-producing decisions (Kanter, 1999; Polanyi 1967).
- Knowledge is a justified personal belief that increases an individual's capacity to take effective action (Alavi and Leidner 1999, 5).
- Knowledge is information made actionable in a way that adds value to the enterprise (1999).
- Knowledge is things that are held to be true in a given context and that drive people to action (Bourdreau and Couillard 1999).
- Knowledge is a "capacity to act" (Sveiby 1997).

Even if we do not want to choose sides in the debate over how knowledge is best defined or what constitutes it, we wish to formulate a definition of knowledge that will be used in the continuation of this paper: *Knowledge is an individual or group capacity developed through formal learning and experience to evaluate and translate data related to the stated problem or objective(s) pursued into meaningful information which enables an effective action.*

This formulation allows us to understand knowledge as an individual and group phenomenon that is intimately linked with action as past experiences influence and present activities. Also it transcends the linear hierarchical division between data, information and knowledge as it defines them as components of a loop - data that become information after evaluation and translation by knowledge, that will become data when transferred (here we take in consideration just explicit knowledge that can be expressed in "hard" form).

8.4 Integral Management Model

In consideration of both the St. Gallen integral management model and Tavcar's model, a new strategic management model was developed (see Figure 1). For the purpose of this study, we want to present only the broad model without inclusion of details (e.g. time differentiation) not relevant to the knowledge management framework development. The central position in the model is occupied by organisation's *vision,* which is the "vector" of stake-holders' interests and management philosophy, and represents the driver for organisation functioning. An organisation's vision can change if the conditions in the environment change to a significant extent or a new dominant coalition is formed. Change in the organisation's vision also means change in other elements of an organisation and they determine how radical and fast the change process can be. *Employees* with the use of tangible and intangible *assets* carry out the organisation's vision through the main *business processes,* such as the supply chain management process, product development management process, and customer relationship management process. Through these processes, inputs are transformed into outputs and added value is created. The amount of the added value is determined by the quality of exchange relationship that exists between an organisation and its market environment. Because every organisation has limited resources and capabilities, it chooses the *business model* (product-market position and value chain) that enables the most efficient use of them *(exploitation)* in line with the organisation's vision and mission. At the same time, through learning *(knowledge exploration),* an organisation develops new competencies or capabilities that will enable it to change the present product-market position or to achieve better results inside it. Within time, employees establish unique interest relationships, which define their roles inside an organisation. Exchange and interest relationships are maintained or changed through organisation processes (management, businesses and support processes) that take place in accordance with formal *operating rules* but first of all in accordance with organization orderliness.

Organisation orderliness attempts to link together organisational culture, structure, management systems, and operating rules in a holistic pattern that brings order to organisation functioning (Biloslavo and Grad 2003). This order is indispensable to balance the natural predilection of an organisation to chaos.

Because organisational culture is very difficult to change when is established, it is an element of normative management (Bleicher 1995; Tavcar 1999). As culture, through shared values, beliefs, and norms, shapes the way people in an organisation interact with each other and with the environment

HORIZONTAL INTEGRATION

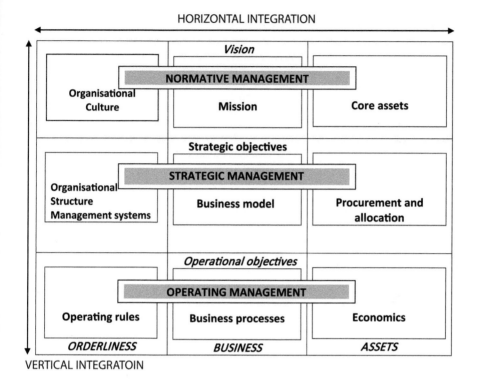

Figure 1 Integral Management Model

outside an organisation, it indirectly has influence on other elements of an organisation. Still this is not a one-way, but a two-way influence where organisation's culture will slowly change if other elements of an organisation will change in a substantial extent (as we know from complexity theory in a dynamic complex system it is not possible to establish cause and effect).

In this way a stability of internal relationships is supported beside organisation culture also with organisation *structure*. Formal structure, which has its horizontal and vertical dimensions, links together different subsystems of an organisation (units and/or functions) that are developed after differentiation and specialisation of some organisation's functions. Harmonious functioning of an organisation is finally secured through *management systems* that pervade all hierarchical and functional levels of an organisation to co-ordinate, direct, activate and control the activities of employees (Viljoen 1994, 458).

8.5 System Knowledge Management Framework

A system is a part of the objective or abstract reality with clear or fuzzy demarcation (e. g. virtual organisation), which is composed of different mutually interconnected elements that are oriented to the achievement of the common purpose. Their relationships conform to the group of system unique rules or/and natural laws. Because some dynamic properties of a system do not exist when a system is decoupled into smaller parts (Rubenstein-Montano et al. 2001, 6) and because to be effective a system requires that its elements fit together, we need a systems approach or systems thinking to understand its behaviour. As Gupta and Govindarajan (2000, 72) said *"the social system should be viewed not as random collection of disparate elements but as a comprehensive whole in which the various elements interact with one another"*

As Figure 2 shows the proposed framework consisting of three strategic building blocks: assets, businesses, and orderliness, that represent both a static and dynamic view of an organisation as well the hard and soft part of it. These elements are permeated with four knowledge management processes that we have identified as knowledge creation, storage, transfer, and application. These processes are based on people and technology and follow one or a mix of both knowledge strategies represented by exploitation and/or exploration of knowledge. Support for the selected elements can be found in the two described strategic management models and different knowledge management frameworks presented in Table 1.

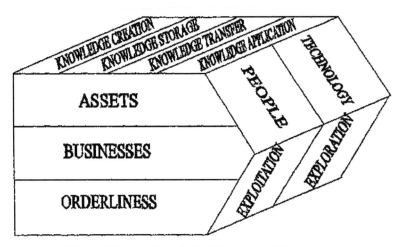

Figure 2 System Knowledge Management Framework

Table 1 Identified Influences on the Knowledge Management

	Culture	Leader-ship	Measurement	Structure	Reward and Incentive Systems	Information and Communication Technology
Alavi and Leidner 2001	✓			✓		✓
Booz Allen and Hamilton 2001	✓	✓			✓	✓
Fouché and Botha 2002	✓	✓	✓	✓		✓
Gold et al. 2001	✓			✓		✓
Mertins et al. 2003	✓	✓	✓		✓	✓
Holsapple and Singh 2001		✓	✓		✓	
Gupta and Govindarajan 2000	✓			✓	✓	✓
Skyrme 1999					✓	
Lai and Chu 2001	✓	✓	✓			✓

8.6 Knowledge Management Processes

The knowledge creation process inside organisation we understand as a dynamic interaction or "generative dance" (Cook and Brown 1999, 393) between knowing and knowledge at the individual and social level, in which new knowledge is generated within the process of learning. This process is composed of four distinctive processes as socialisation, exteraalisation, combination, and internalisation (Nonaka and Takeuchi 1995) that take place inside a micro-community or communities of practices. New knowledge that is created in the knowledge creation process needs to be store for later used as an organisational memory. The processes of knowledge storage involves finding ways to convert documents, models, human insights and other artefacts into forms that make retrieval and transfer easy without losing the "true meaning" of the knowledge (Staples et al. 2001, 11). With the use of information technology, organisations try to-develop vast repositories of organisation knowledge about customers, projects, processes, suppliers, competition, technology, industry and organisation's knowledge itself that can be retrieved or transfer at any time anywhere. Knowledge transfer occurs at various levels of an organisation, for example between individuals, between individuals and groups, between groups, between groups and an organisation, and between organisations (Alavi and Leidner 2001, 119). If we consider knowledge as an independent phenomena from the context where it is produced or used, then we can say that an organisation must try to transfer the right knowledge at the right time to the locations where it is needed.[1] This process can be supported mostly by information and communication technology as in an organisation

---

[1]In consideration of our definition of knowledge what can be transferred are data and not knowledge. Data can become knowledge in consideration of context and existing knowledge of the receiver (knowledge absorption capacity). Here we use the notion of knowledge just to be in line with a major part of the literature.

- Capability to learn: The employee should know what, how and where to learn best. Here we do not consider just single loop learning where people perform differently without changing their mental models, but also double loop learning that manifests itself when people are capable of reconsidering their own mental models (Argyris and Schon 1978).

- Task competencies: These refer to the job task competence, as well as the capability to utilise new technologies.

- Communicative competence: The employee should have the ability to express him/herself and to be able to acquire, interpret and filter information,

- Flexibility: The employee should be capable of thinking outside mainstream framework and to solve problems.

that uses a codification strategy or by an extensive personal networks as in an organisation that uses a personalization strategy (Hansen et all 999). Without knowledge application, all the aforementioned processes are useless. Only knowledge application can ensure that the organisation knowledge represents a viable source of competitive advantage. To be of value for organisation's stakeholders disposable knowledge needs to be transformed in a lower cost structure, a larger revenue stream or both.

8.7 People and Technology

Two basic elements of knowledge management are people and information and communication technology. Both of them often represent the basis of argument between so-called technology and human-oriented researchers. However, in our framework we consider both to be of equal importance. The reason for this stance is that knowledge is inseparably linked to people, therefore an organisation cannot create new knowledge without them. On the other hand, an organisation cannot efficiently use disposable knowledge without the right technology.

8.7.1 People

According to Churchman (1972) knowledge resides in the user and not in the collection of data; therefore an organisation needs to focus its knowledge management effort not on data, but on its people. This task is even more difficult if we consider that people are not only the key enablers in creating and using knowledge for competitive advantage, but they are also the major constraints.

The literature review focused attention on three different attributes of people as carriers of organisational human capital:

1. *Leadership* as a capability to develop a clear vision of the present and future organisation's needs for knowledge and being able to motivate people to learn and innovate.
2. *Adaptability* as a capability of people to be aware of changes in the outside world and to be prepared and competent to deal with them.

- Social competence and team capabilities: The employee should know how to negotiate effectively within groups and how to share task accomplishments.
- Autonomy: The employee should identify him/herself with an organisation and not just as an order recipient.

3. *Networking* as a capability of people to build and sustain a social network of colleagues and professional acquaintances that supports knowledge creation and sharing.

The literature supports the idea of leadership as one of the critical factors for effective knowledge management (Rastogi 2000; Kirrane 1999; Davenport, De Long and Beers 1998; Hasanali 2002; Halal 1998; Holsapple and Singh 2001; Skyrme 1999). The leadership of the organisation sustains the vision and mission of an organisation and supports organisational values by leading by example. Knowledge leaders encourage communication and collaboration; recognise and reward good ideas and innovations; put high emphasis on training and learning, and install performance-based promotion system to motivate people and build trust (Davenport, De Long and Beers 1998; Halal 1998; Skyrme 1999). Only leadership that understand that effective knowledge management can represent a source of competitive advantage in its own right and values trust among employees can build a winning organisation that knows how to balance the need for new knowledge with a need for efficient use of disposable knowledge.

Because knowledge can quickly become irrelevant in light of changes external and internal to the organisation, it is critical that individuals, who have unique bundles of competencies, can adapt their competencies to new situations as rapidly as possible. Employees' competencies that can be considered as essential in the conception of adaptability are (Boyett and Conn 1992; Katzenbach and Smith 1994, 47):

As communication theory suggests a network's potential benefit grow exponentially as the number of nodes expand (Evans 2003, 82). People that are capable to build a network can uncover more opportunities for knowledge creation and application and at the same time contribute to build trust inside an organisation. This capability is important in every organisation, but especially in a multinational global organisation where it can contrast the possible negative effect of cultural diversity. Within the whole knowledge creation process the ability to build a social network is not confined only to a socialisation process. It also plays a prominent role in justifying a concept when a micro-community or creative individuals need to persuade decision makers within an organisation to support the next phases of their work.

8.7.2 Information and communication technology

Information and communication technology (ICT) has made it possible to preserve valuable explicit knowledge for the future and to share a huge amount of information unconstrained by the boundaries of geography and time. For an organisation, this means an opportunity to horizontally and vertically integrate task and data and in this way to shorten the length of the transformation cycle. The transformation cycle includes not only transformation of tangible inputs into products/services but also transformation of intangible ideas and insights into tangible outputs. With regard to Hamel and Prahalad's (1995) claim that it is not the absolute level of knowledge a firm possesses which leads to competitive advantage, but the velocity with which it is circulated in the organisation, we can understand why ICT is so critical for knowledge management success in an organisation.

We divided information and communication technology into three groups:

1. *Technology for knowledge codification and storage* that includes different types of knowledge repositories and knowledge-based decision support systems. Software agents which can search for information across many repositories on behalf of the user and data mining tools that help to identify new patterns in large volumes of data also belong to this kind of technology.
2. *Communication technology* that supports knowledge transfer irrespective of its format, user operating system, or communication protocols. Communication technology also includes knowledge maps, which are pointers to knowledge providers inside or outside an organisation.
3. *Collaborative technology* that enhances person-to-person collaboration which can happen at the same or different time and in the same or different place.

Which of these technologies is the most important for an organisation depends on the organisation's context shaped by orderliness, businesses, and disposable assets as well as on the organisation's knowledge strategy. Anyway an effective virtual ba (Nonaka et al. 2003, 499) demands a congruent combination of all of them.

8.8 Exploitation and Exploration Knowledge Strategies

Several issues must be considered in relation to an organisation's knowledge strategy direction. The first thing is that an organisation's decision to exploit disposable knowledge or to explore for new knowledge significantly impacts the type of knowledge that an organisation needs, the way knowledge is used, and the design of organisation's constitutive elements. To assess which strategy or mix of both strategies is better for an organisation's long-term success, the knowledge content of organisation's core assets must be identified and compared with the need of an organisation's vision and with competitors' knowledge. In this way, it is possible to estimate the knowledge gap that exists between what an organisation already knows and what it needs to know to exploit opportunities and avoid threats in the marketplace.

If an organisation is mostly oriented to exploitation of its internal knowledge then organisational culture is predominantly inside oriented with high use of information and communication technology to support internal knowledge creation, store, transfer, and application. Incentives are used to reward employees who efficiently exploit the organisation knowledge base and links that are part of a "hidden" layer of structure stay inside an organisation. Leadership supports efficiency and fitness between organisational and individual values is an essential internal characteristic. On the other hand, an organisation that is mostly oriented to exploration of external knowledge has an externally oriented culture where a "hidden" layer of structure is developed outside organisational boundaries. Information and communication technology is first of all used for identifying and solving new or unique problems. With leadership that supports innovations, new knowledge creation and transfer of outside knowledge into an organisation are rewarded. A key characteristic of organisation's employee is adaptability.

8.9 Normative Management

As previously discussed, normative management represents the long-term framework that gives direction to organisation's activities. Normative management results from an organisation's vision that defines what kind of knowledge the organisation should create and in what domain (Nonaka et al. 2003, 506) or for what it must strive for in the long run. Without a clear knowledge vision an organisation is not in a position to align its constitutive elements and knowledge management activities, or to focus its knowledge management efforts. As a result, any action taken in consideration of an organisation's knowledge base development is almost certainly condemned to fail.

Normative management consists of:

1. *Core knowledge assets* that consist of experiential, market, and systemic knowledge assets (adapted after Nonaka et al. 2003, 502).

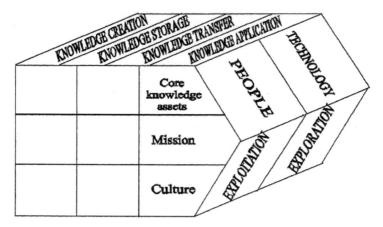

Figure 3 Elements of Normative Management in System Knowledge Management Framework

2. *Organisation mission* that defines to who and how an organisation will serve.
3. *Organisation culture* that represents a tacit part of group knowledge developed through common shared expe.

8.10 Core Knowledge Assets

Core knowledge assets are a totality of distinctive tacit and explicit knowledge at an individual and/or organisational level that represents the present source of organisation's competitive advantage, with the exception of tacit knowledge embedded in actions and practices as organisational culture and operating rules.

Organisation's core knowledge assets consist of (adapted after Nonaka et al. 2003, 501–503);

- *Experiential knowledge assets* that include tacit knowledge shared through common experiences as skills and know-how of individuals that are organisation-specific.
- *Market knowledge assets* that Include explicit and tacit knowledge articulated through images, symbols, and language and embedded in product concepts, design and brand equity.
- *Systemic knowledge assets* that include codified explicit knowledge embedded into documents, manuals, databases, patents and licenses.

Core knowledge assets are elements of the normative management because represent the starting point for formulation of the organisation's strategy. On the other hand they also represent the core rigidities of an organisation as formulated by Leonard-Barton (1995). For both reasons an organisation needs to identify them as to be able to exploit them for today's success and to change and improve them for future success through new knowledge creation.

8.11 Organisation Mission

Organisation mission is a statement about sense of the organisation's existence. It defines a broad direction that an organisation wants to follow in consideration of its stakeholders and a broad framework for its market oriented activities.

Organisation mission is composed of the answers on three main questions:

- *To whom* do we as an organisation serve?
- What are interests that we as an organisation want to realise?
- *How* will we realise the added value on the market?

The first two answers are intimately linked with organisation culture, while the last one with core knowledge assets. An organisation mission is important (1) because it defines what knowledge will be important for an organisation in consideration of its stakeholders and their needs and (2) because it indirectly express what knowledge an organisation has about own identity and its environment. In this way an organisation mission directs a sense-making process inside an organisation (Weick 1995).

8.12 Organisation Culture

Because organisational culture determines the kinds of knowledge sought and nurtured, and the kinds of knowledge-building activities tolerated and encouraged within an organisation (Leonard-Barton 1995), a significant proportion of the literature that was reviewed considers organisation culture as "the key" factor for success of knowledge management initiatives (Martin 2000 24; De Long and Fahey 2000; Rastogi 2000; Bock 1999; Holowetzki 2002). From a knowledge management point of view, a key issue for an organisation is to instil an organisation wide culture that encourages change and openness to new knowledge regardless of where this knowledge is developed.

Three major sub-factors of organisation culture that can be found in the literature are:

- *Readiness for risk taking* where risk and failures are recognised as organisational and individual learning.
- *Knowledge sharing* with a clear understanding of the mutual benefit in doing it and aclimate of openness and trust.
- *Outside orientation* as represented by the urge to exploit and develop knowledge presented in collaborative joint ventures, alliances and partnerships.

To promote the learning process and new knowledge creation an organisation needs to sustain a culture where calculated risk of failures is acceptable and expected, creative solutions are always considered in addition to more conventional ones, and people are given time and resources to try new things. Readiness for risk taking changes in different contexts, but as a culture element it strongly differentiates innovative and visionary organisations from status quo organisations. Without a culture of knowledge sharing, knowledge remains the property of an individual or a group inside an organisation and not available to the larger organization. Such a culture does not promote communication between employees or reward the exchange of knowledge. In this way distrust between employees develops and an attitude where people are primarily concerned about their own benefits before anything else.

Outside or external orientation represents an attempt to go beyond traditional organisational structures and boundaries in favour of the establishment of symbiotic arrangements with external partners, which can provide a mutual advantage (Wigand et al. 1997, 209). Two reasons exist why outside oriented cultures are not often found in the organisations. The first is that an organisation may be unsure if it will be able to protect its own knowledge from leaking. The second reason, is an intimate fear to concede that something has changed outside an organisation in a way that an organisation is no longer the best or the only one in consideration of the matter involved ("not-invented here" syndrome).

8.13 Strategic Management

Strategic management represents the link between the long-term normative management and short-term operating management. If the organisation's vision represents an absolute or ideal standard that an organisation wants to achieve in the future and the operating management represents a historical standard that an organisation needs to improve if it wants to survive in the middle and long-term, then the strategic management represents an external

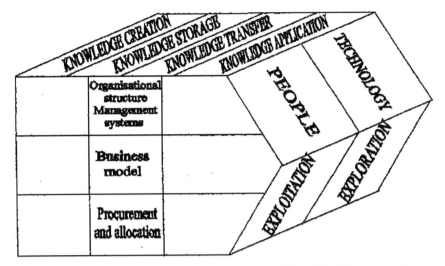

Figure 4 Elements of Strategic Management in System Knowledge Management Framework

standard of comparison with other organisations. The main goal of strategic management is to achieve and maintain the competitive advantage of an organisation.

8.14 Organisation Structure

Organisational structure is in some way a result of organisational culture. However, as we stated earlier, the influence between these is not one-way and a restrictive structure could inhibit a knowledge enriching culture.

After Fouche and Botha (2002) we divided structure of an organisation into the three layers of "network forming devices":

1. *Formal hierarchical structure* with reporting relationships, responsibilities and accountabilities.
2. *Flexible structure* that consists of ad hoc problem-solving teams, task forces, joint planning groups etc.
3. *"Hidden" (implicit) structure* that consists of informal peer groups, interest groups, professional groups and personal networks internal as well as external to the organisation.

In the literature, considerable evidence can be found that a formal hierarchical structure prevents effective knowledge management (De Long and

Seeman 2000). An organisational structure that does not preclude communication and collaboration across hierarchical and functional boundaries must be permeable in the sense of a free flow of knowledge between employees regardless their job position, job function, or any other traditional boundaries (Symon 2000).

Because formal structure can inhibit knowledge flow in an organisation, it is very important how effectively the organisation employs flexible and hidden structures to facilitate the flow of knowledge. A flexible structure is composed of project teams and other task-oriented groups that can accommodate multi-disciplinary and cross-functional members. The role of a flexible structure is not just to promote more flexibility of the organisation, but also to motivate employees to develop and share their knowledge. Hidden structures promote the exploitation of opportunities created by a workplace setting of open spaces where co-location and informal meeting places are part of daily organisational routine. This layer of structure represents the biggest share of the "social capital" of an organisation as a particular combination of networks that exist inside and outside an organisation. Hidden structure have achieved more popularity in the last few years with developed interest in communities of practice, which represent informal groups that interacted and collaborated regularly around work-related issues and challenges (Cross and Baird 2000). With the focus on communities of practice that exist in an organisation, it is possible for an organisation to promote knowledge sharing and creation, and in this way to secure a higher possibility of success.

8.15 Organisation Systems

The purpose of organisation systems in relation to knowledge management is to enable effective and efficient application of the knowledge within an organisation. For promoting knowledge management in the organisation we need a performance system that uses tangible (financial) and intangible (non-financial) long-term oriented measures and a reward system that incorporates both extrinsic and intrinsic rewards (Davenports De Long and Beers 1998). As contingency theory suggests, coherency between different kinds of organisation's systems is absolutely critical for success of any knowledge management initiatives.

Organisation systems are composed of:

1. *Reward and recognition system* that needs to be broad enough to encourage people to learn and share knowledge with their co-workers

and access knowledge that is available in organisation information system.

2. *Planning and performance system* that includes organisation's objectives and measures for both tangible and intangible aspects of an organization.

3. *Information system* that supports store and transfer of data from outside or inside of an organisation and has clear rules for categorizing organisation's products and processes.

Because the possession of knowledge is thought to give a personal competitive advantage (De Long and Fahey 2000, 113), it is difficult without the right reward and recognition system to facilitate the sharing of knowledge between an individual - employee and an organisation. For this reason, knowledge management literature often emphasized that an organisational reward and recognition system needs to be made broader, encouraging people not to compete with one another but to share knowledge and to work together for creation of new knowledge (Habbel et al., 1998).

We believe an effective reward and recognition system needs to make equal use of extrinsic (tangible) and intrinsic (intangible) rewards. Extrinsic motivators are above all oriented to an assessment of achievement against knowledge management objectives such as (1) acquiring new knowledge; (2) undertaking new projects or responsibilities; (3) contributing to a community or team; (4) contributing to the development of another employee (Brelade and Harman 2000), On the other hand, intrinsic motivators are based on more informal, short-term rewards that give employees a feeling of accomplishment with regard to knowledge development, creation or sharing. If we consider that intrinsic motivators lead to solutions that are far more creative than do extrinsic motivators, and the former are the most immediately affected by the environment (Amabile 1998), we can conclude that in this area, a knowledge oriented organisation can make an significant positive difference compared to competitors.

In the last decade or more, attention has been oriented to the organisation's internal resources and capabilities and a major shift from top-down planning to bottom-up and middle-up-down strategy development has occurred (Nonaka and Takeuchi 1995; Floyd and Wooldridge 2000). Beside these changes that try to build and cultivate a more collaborative and participative environment inside an organisation, important changes have occurred with regard to performance measurement systems. These are now more future oriented and it try to take into account the intangible assets of an organisation such as knowledge. The literature already mentions different performance measurement systems which consider and measure intangible assets of organisation, like Skandia Navigator (Edvinsson and Malone 1997), Balanced Scorecard (Kaplan and Norton 1996), Intangible Asset Monitor (Sveiby 1997), IC-Index (Roos et al. 1997) etc. Despite these and other systems, we do not currently have a complete system for the measurement of intellectual or intangible assets (Bontis 2001). However, in consideration of a "paradigm shift" (Bontis 2001) it can be supposed that with the further development of the above-mentioned systems or with the development of new ones, organisations will eventually have a tool, to help better manage their core knowledge assets. Information system encompasses much more than technology for facilitating data store and transfer. An organisation's information system is composed of technology, people and processes that need to be put together consistently if an organisation wants that the system is capable (1) of helping people inside an organisation to identify problems and opportunities, and (2) of storing and transferring data that reside inside or outside an organisation in the way that ensure the optimal "intelligence density" (Gallupe 2001). In addition an effective information system identifies obsolete data and replaces it.

8.16 Business Model

In consideration of the broader organisation's mission, the organisation's business model represent an explicit answer to three questions (Govindarajan and Gupta 2001, 3):

- Who are the organisation's target customers?
- What value do we as an organisation want to deliver to them?
- How will the organisation will create it?

As consequence of the above-mentioned questions, we can say that the business model involves three main elements (Govindarajan and Gupta 2001, 4):

- *Customer base* that defines market segments where the organisation can exploit its competitive advantage.
- *Concept of customer value* that it goes beyond "hard" product or service to intangibles with a high proportion of knowledge transferred from the organisation to the customers.

- *Value chain architecture* that represents chain of activities that an organisation needs to perform to deliver value to the customers in the most efficient way.

The business model is the end result of the organisational intelligence activities linked with the organisation's vision. In the process of business model development, an organisation uses its knowledge about the marketplace and its future development together with its knowledge about its own core knowledge assets. The knowledge that it uses is predominantly implicit in nature, especially if an organisation tries to change the existing rules of the game. How this knowledge is created, stored, transferred and applied will significantly impact the business model that an organisation will develop.

8.17 Procurement and Allocation

To successfully procure and allocate "knowledge assets" an organisation needs to have a clear understanding of its disposable core knowledge assets and the requirements of its business model. Only after an assessment of the gap that exists between them will an organisation be in a position from which it can decide where and how it will procure and allocate needed assets.

Procurement and allocation as an element of the strategic management needs to consider three types of "knowledge assets":

- *Knowledgeable people - experts* represent individuals outside the organisation that can bring important knowledge to an organisation if they are involved in its operation.
- *Knowledgeable products or services* represent products or services that an organisation atthe moment for different reasons is not able to produce or developed, but if they are found in the marketplace at the right price and properly employed they can have a strong impact on competitive advantage of an organization.
- *Knowledgeable organisations* represent competitors and other organisations that hold knowledge assets that an organisation does not have and it can not develop without incurring in disproportional costs.

When an organisation considers the acquisition of "outside" experts, it needs to think about some of the negative consequences that such a decision might bring:

- People inside the organisation (especially middle managers) can become disinterested in upgrading competencies,

- Trust and loyalty inside the organisation can decrease,
- The organisation can loose an opportunity to capitalise on the knowledge of its veteran employees,
- Newcomers can bring new knowledge, but they can also make serious errors because they are unfamiliar with the way to behave and work inside the organization.

For these and other reasons is very important that an organisation tries to balance promotion and development of its existing employees with the acquisition of experts from outside. It also needs to consider whether or not the expert fits with its core values, and if the knowledge really resides in the expert or in the environment where he/she works.

Knowledgeable products or services can often represent a unique opportunity for an organisation strengthen its future sources of competitive advantage. To achieve this an organisation needs to take a proactive stance. This requires that it attempts to not only improve the acquired product or service, but to develop or at least contribute to the development of the second generation of it. For a lot of small and middle sized organisations this represents the only way for development and upgrading of its core knowledge assets.

With respect to knowledgeable organisations, we can say that an organisation has two main approaches it may use. The first is to acquire another organisation, the second is to build a partnership. Organisations need to decide which approach to use by considering the of different internal and external contexts.

8.18 Operating Management

Operating management represent the front stage where an organization produces and delivers value to its customers.

Sadly, operating management is often neglected in the knowledge management literature with the exception of business processes. In our developed framework we also wish to contribute to a consideration of this level of management within the knowledge management literature.

8.19 Operating Rules

As Eisenhardt and Sull (2001) pointed out well defined simple rules are essential for an organisation's success in today's unpredictable world. The optimal number of rules that an organisation uses can change over time as the business landscape changes (Eisenhardt and Sull 2001, 113). In a period of predictability an organisation uses more rules in order to increase efficiency. In a period of turbulence it uses less rules in order to increase flexibility.

After Eisenhardt and Sull (2001) operating rules are divided into three types:

- *Priority rules* help managers rank opportunities and problems that they encounter in day-to-day activities.
- *Timing rules* help managers to synchronise their activities and decisions with other parts of an organisation,
- *Exit rules* help managers to decide when is the right time to finish with the activities or projects that do not have prospects of success.

Operating rules are very rarely the result of careful thinking and analysis, more often they are results of experience, especially of painful ones. They are products of collective problem solving where organisational culture develops as social tacit knowledge and operating rules as social explicit knowledge.

8.20 Business Processes

Organisations not only create knowledge, they also, and usually primarily, create goods and services. In accordance with Rastogi (2002, 234): "the test of value creation is whether customers are willing to pay for a firm's product(s) and or service(s) under conditions of competitive choices open to them," Knowledge management initiatives that do not consider the utility of knowledge for value creation are from an economical point of view completely useless and just destroy valuable organisational resources.

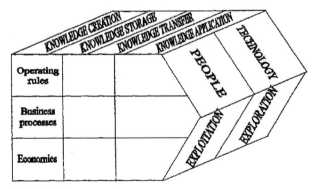

Figure 5 Elements of Operating Management in System Knowledge Management Framework

Operation level value activities are executed through business processes as an interrelated, sequential set of activities and tasks that turn inputs into outputs, and have a, distinct beginning, a clear deliverable at the end and a set of metrics that are useful to, measure performance (Pearison 2001).

After Fahey et al. (2001, 894) business processes are divided in three main groups:

1. *Product development management process* that is oriented to the development of new customer solutions and/or to the improvement of existing solutions.
2. *Supply chain management process* which purpose is the acquisition of inputs and their transformation into desired customer benefits.
3. Customer *relationship management process* that develops and nurtures relationships with external marketplace entities.

Knowledge integrates all three business processes into a coherent whole and adjusts them in response to external change in order to better fulfill customers' needs. This can be achieved by transforming marketplace data into knowledge, by transforming primarily products or services to knowledge-based ones, by changing a buy-sell attitude into a customer-supplier relationship, by changing the focus from an individual process to share understanding of a business model, and by reconsidering an organisational value chain as a dynamic value net.

8.21 Economics

In general an organisation can be oriented to exploit one of the three possible economies:

- Economies of scale exist when an organisation can reduce the unit cost by highproduction volumes.
- *Economies of scope* exist where the same equipment can produce multiple products morecheaply in combination than separately (Goldhar and Jelinek 1983).
- *Economies of substitution* exists when the cost of designing a new system through the partial retention of existing components is lower than the cost of designing the system afresh (Garud and Kumaraswamy 2002).

Economies of scale are the product of the industrial revolution, in the same way as economies of substitution are a product of information revolution. The type of economies that an organisation has chosen to exploit will direct type of knowledge that it will develop and use.

8.22 Conclusion

For an understanding of relationships that exist between knowledge management processes and organisation elements and their impact on an organisation's success we need more than just a list of key elements. What we require is an identification of relevant sub-dimensions of broad organisational elements and their analysis, to get a clearer picture of an organisation as purposeful system of knowledge creation and exploitation. This paper presents the first phase of our research. In the future this research will be complemented with survey and analytical analysis of my hypothesis so as to discover how the proposed framework can be useful for management practice, especially for knowledge management audit and assessment of the impact that knowledge management processes and organisational elements can have on an organisation's success. Hopefully this framework can provide a right basis for future research and further refinement of identified elements.

References

1 Alavi, M., and D. E. Leidner. 1999. Knowledge Management Systems: Issues, Challenges and Benefits. Communications of the Association for Information System 1 (5).

2 Alavi M., and D. E. Leidner. 2001. Review: Knowledge Management and Knowledge Management System: Conceptual foundations and research issues, MIS Quarterly 25 (1): 107–136.

3 Amabile, T. M. 1998. How to Kill Creativity. Harvard BusinessReview 76 (5); 76–88.

4 Argyris, C., and D. A. Schon. 1978. Organisational Learning: A Theory of Action Reading: Perspective, Addison-Wesley.

5 Bartlett, C. A., and S, Ghoshal. 1994. Changing the Rule of Top Management: Beyond Strategy to Purpose. Harvard Business Review (Nov.–Dec.): 79–88.

6 Biloslavo, R., and 1 Grad. 2003. Fuzzy Expert System for Evaluating the Flexibility of an Organization: Theoretical Foundations and Field Study Research. Issues in information Systems 1: 25–31.

7 Bleicher, K. 1995. Das Kozept integriertes Management. Frankfort and New York: Campus Verlag.

8 Bock, F. 1999. Viewing K M in Terms of Content, Culture, Process and In-frastructure:The Intelligent Approach to Knowledge Management. Knowl-edge Management Review 2(I): 22.

9 Bontis, N. 2001. Assessing Knowledge Assets: A Review of the Models Used to Measure Intellectual Capital International Journal of Management Review 3 (1): 41.

10 Bourdreau, A., and G Couillard. 1999. System Integration and Knowledge Management. Information System Management 16 (4): 24.

11 Boyett, J, H., and H. P. Conn. 1992: Workplace 2000: The Revolution Reshaping American Business, New York: Penguin.

12 Brelade S., and C Harman, 2000. Using Human Resources to put Knowl-edge to Work. Knowledge Management Review 3(1): 26.

13 Churchman, C W, 1972. The Design of Inquiring System: Basic Concepts of System and Organizational New York: Basic Books,

14 Cook, S. D. N., and J. S. Brown. 1999. Bridging Epistemologies: The Generative Dance between Organizational Knowledge and Organizational Knowing. Organizational Science 10 (4): 381–400.

15 Davenport, T. H, and L. Prasak. 1998. Working Knowledge: How Or-ganization Manage What They Know. Boston: Harvard Business School Press.

16 Davenport, T. H., D,W. De Long, and M. C. Beers. 1998. Successful Knowledge Management Projects, Sloan Management Review 39 (2): 43–57.

17 De Long D. W., and P, Seeman 2000. Confronting Conceptual Confusion and Conflict in Knowledge Management, Organizational Dynamics 29 (!): 33–44.

18 De Long, D. W., and, L. Fahey, 2000. Diagnosing Cultural barriers to knowledge management. Academy of management Bxecu$ivel4 (4): 113.

19 Eisenhardt, L,; and M. S. Malone. 1997. Capital: Tour Company's True Value by Finding Us Hidden Brainpower, New York: Harper Business.

20 Eiseohanft, K. M., and D. N. Sull 2001, Strategy as a simple rule. Harvard Review 79 (January): 107–116.

21 Evans, C. 2003. Managing for Knowledge: HR's strategic role, Amster-dam; Butterworth Heinemann.

22 Fahey, L et al 2001. Linking E-business and Operating Processes: The Role of Knowledge Management. IBM System Journal 40 (4); 889.

23 Floyd, S.W., and B. Wooldridge. 2000: Building strategy from the middle: Reconceptualizing Strategy Process. Thousand Oaks: Sage.

24 Fouche, B., and D. F. Botha. 2002. Knowledge Management Practices in the South African Business Sector: Preliminary Findings of Longitudinal Study. South African Journal of Business Management 33 (2): 13.

25 Gallupe, B. 2001. Knowledge Management System: Surveying the Landscape. International Journal of management Reviewers 3 (1): 61.

26 Garud, R., and A. Kumaraswamy. 2002. Technological and Organizational Designs for Realizing Economies of Substitution. In The Strategic Management of Intellectual Capital and Organizational Knowledge^ edited by Choo, C, W., and N. Bontis, Oxford: Oxford University Press.

27 Goldhar JL D., and M. Jelinek. 1983. Plan of Economies of Scope. Harvard Business Review (nov.- dec.): 141–148.

28 Gomez, P. 1999; Integrated Value Management London: International Thomson Business Press.

29 Gomez, P., and T. Zimmermann. 1993. Unternehmensorganisation — Profile, Dynamik, Methodik Frankfurt, New York: Campus Verlag.

30 Govindarajan, V., and A, K. Gupta 2001. Strategic innovation: A conceptual Road Map. Business Horizons (July- August): 3–12.

31 Greenwood, R., and Hinings, CR., 1993: Understanding Strategic Change: The Contribution of Archetypes. Academy of management Journal 36: 1052–1081.

32 Gupta, A, K., and V. Govindarajan. 2000. Knowledge management's Social Dimension: Lessons From Nucor Steel. Sloan Management Review. Fall 71–80.

33 Habbel et al. 1998, Knowledge management: knowledge-critical capital of modern organizations, Booz Allen & Hamilton Insights, [www.bah.com/ viewpoints/insights/cmt_knowledge_2.html]

34 Halal, W. E. 1998. Organizational Intelligence; What is it, and How Can Managers Use it to Improve Performance? Knowledge Review 1 (March–april); 20–25.

35 Hamel, G., and C. K. Prahalad. 1995. Competing for the Future, Boston; Harvard Business School.

36 Hansen, M. T. et al. 1999. What's Your Strategy for Managing Knowledge? Harvard Business Review, 77 (2): 106–117.

37 Hasanali, F. 2002. Critical Success Factors of Knowledge Management http://www.kmadvantage.coni/dQcs/km_articles/Critical^Success^Factore_ofJKM.pdf

38 Holsapple and, M. Singh. 2001. The knowledge chain model: Activities for Competitivnes. Expert Systems -with Applications 20 (l):77–98.

39 Holsapple, C.W., and K.D Joshi 1999. Description and Analysis of Existing Knowledge Management Frameworks^ Proceedings of the 32nd Hawaii International Conference on Systems Science, 1999,

40 Kanter, J, 1999. Knowledge Management, Practically Speaking. Information Systems Management 16 (4): 7.

41 Kaplan R. S. and D. P. Norton 1996 The Balanced Scorecard: Translating Strategy Into Action. Boston: Harvard Business School Press.

42 Katzenbach, J. R., and D. K. Smith. 1994. The Wisdom of Teams. New York: Harper Business.

43 Kirrane, D. E. 1999. Getting wise to knowledge management. Association Management 51 (8): 31.

44 Leonard - Barton, D. 1995. Wellsprings of Knowledge: Building and Sustaining the Sources of Innovation. Boston: Harvard Business School Press.

45 Manz, C. C. and H. P. Sims. 1993. Business without Bosses: How Self-Managing Teams are Building High Performing Companies, New York: John Wiley & Sons.

46 Martin, B. 2000. Knowledge Management within the context of management: A evolving relationship. Singapore Management Review 22 (2): 17–36.

47 McDermott., R. 1999. Why Information Technology Inspired but Cannot Deliver Knowledge Management, California Management Review 41 (4): 103–117.

48 Mertins, K. etal. 2003. Knowledge Management: Concepts and best Practices, 2nd ed. Berlin: Springer,

49 Meyer, C. 1993'. Fast cycle time: How to align purpose, strategy, and structure for speed. New York: Free Press.

50 Nonaka, L et al. 2003. A Theory of Organizational Knowledge Creation: Understanding the Dynamic Process, of Creating Knowledge, In Handbook of Organizational Learning and Knowledge, edited by. Dlerkes, M., A, Berthoin-Antal, J. Child and I.Oxford: Oxford University Press.

51 Nonaka, L, and H. Takeuchi 1995. The Knowledge Creating Company. New York; Oxford University Press.

52 Polanyi, M. 1967. The Tacit Dimension. London: Routledge and Keoan Paul

53 Rastogi, P, N. 2000. Knowledge Management and Intellectual Capital - The New Virtuous Reality of Competitiveness, Human Systems Management 19 (1); 39.

54 Rastogi, P. N. 2002. "Knowledge Management and Intellectual Capital as a Paradigm of Value Creation", in; Human Systems Management, V. 21, 1. 4, 229–240.

55 Roos, J. et al. 1997. Intellectual Capital. Basingstoke; Macmillan Business.

56 Rubenstein-Montano, B. et al. 2001. A systems thinking framework for knowledge management. Decision Support Systems 31 (1): 5.

57 Schein, E, H. 1985. Organisational Culture and Leadership. San Francisco; Jossey-Bass Publishers.

58 Schwaninger, M. 2001. System Theory and Cybernetics: A Solid Basis for Traesdisciplinary in Management Education and Research. Kybernets 30 (9/10); 1209–1222.,

59 Schwaninger, M. 1994. Managements system, Frankfurt, New York; Campus Verlag.

60 Skyrme, D. J. 1999. Knowledge Networking: Creating the Collaborative Enterprise. Woburn: Butterworth-Heinemann.

63 Staples, D, et al. 2001. Opportunities for research about managing the knowledge-based enterprise. International Journal of Management Reviews 3(1); 20.

64 Sullivan, G. R., and M. V. Harper. 1996. Hope is not a Method: What Business Leaders Can Learn from America '$ Army, New York; Broadway Books.

65 Sveiby, K. E, 1997. The New Organizational Wealth: Managing Measuring Knowledge-based assets. San Francisco; Berret-Koehler.

66 Symon, G. 2000, Information and Communication Technologies and the Network Organization; A Critical Analysis. Journal of Occupational & Organizational Psychology, 73 (4).

67 Tavcar, M, I. 1999. Razseznosti $trateskega management, 2 izdaja. Koper: Visoka Sola za management.

68 Tavcar; M, I, 2003. Strateski management Koper; Visoka Sola za management.

69 Vail, III E. F. 1999. Knowledge Mapping; Getting Started With Knowledge Management Information Systems 16 (4); 16,

70 Viljoen, J. 1994. Strategic Management: Planning and implementing successful corporate strategies. Melbourn:Longman.

71 Weick, K. E, 1995. Sensemaking in Organizations. Thousand Oaks: Sange.

72 Wigand, R. et al. 1997. Information, Organisation and Management - Expanding marketsand Corporate Boundaries. Chichester: John Wiley & Sons.

9

HR Practices, Social Climate, and Knowledge Flows: Towards Social Resources Management

Angelos Alexopoulos[1] and Kathy Monks[2]

[1]*Europion Council for Nuclear Research Switzerland*
[2]*The learning Innovation and Knowledge Research centre,*
DCU Business School, Dublin City University, Glasnevin,
Dublin 9, Ireland

9.1 Abstract

Despite theoretical support suggesting a strong linkage between HR systems and knowledge management outcomes, only limited empirical evidence exists on the relative contribution of HR practices, particularly as experienced by individual employees, to facilitating intrafirm knowledge flows. Further, even fewer studies have investigated key intermediate mechanisms by which HR practices affect knowledge sharing attitudes and behaviour. Drawing on a survey of 135 core knowledge employees from three Irish-based firms, we found that reciprocal task interdependence, feedback from others, selective staffing and socialisation, relationship-oriented training and development, and line management support for knowledge sharing were the main factors associated positively with employee perceptions of a Social climate that encourages cooperation and teamwork orientation. The implications of our findings are discussed.

Keywords: Human resource management, Knowledge work, Social climate, Cooperation, Knowledge sharing.

9.2 Introduction

In parallel with the widespread recognition that the transfer of people-embodied knowledge is a core basis for competitive advantage available in firing (Argote & Ingrain, 2000), attention has recently focused on the role of the HR function in advancing the knowledge and knowing capability of the firm and, consequently, its value proposition (e.g., (e.g., Storey & Quintas, 2001; Jackson, Hitt & DeNisi, 2003; Kang, Morris, & Snell 2007; Svetlik & Stavrou- Costea, 2007). While human resource management (HRM) scholars are increasingly aware of the importance of fit between knowledge management (KM) initiatives and people-related issues, there are significant gaps in understanding the synergies between HR practices and KM processes and outcomes. In particular, the mechanisms through which HR practices affect employee attitudes and behaviour towards participating in knowledge sharing activities remains a largely unresolved question. The objective of this paper is to address this gap by arguing that HR practices may influence employee knowledge sharing attitudes and behaviour through their impact on perceptions of an organisational social climate conducive to cooperative social relations and teamwork orientation. Such a climate has been identified in the literature as key to knowledge exchange and organisational learning (Nahapiet, Gratton, & Rocha, 2005; Jackson, Chuang, Harden, & Jiang, 2006). The key theoretical contribution of this article lies in nudging the dialogue on the HRM-knowledge-performance linkage from human capital to social relations. HRM research has traditionally focused on methods of developing human rather than social capital (Brass & Labianca, 1999; Leana & Van Buren, 1999). From an individualistic HRM perspective, the social climate of the firm is considered little more than a context for individual needs, interests, values, motivation, and behaviour (Brass, 1995). However, given that the firm's knowledge and knowing capability depends both on human and social capital advantage, 'to focus on the individual in isolation is, at best, failing to see the entire picture. (Brass & Labianca, 1999; 323). This article seeks to bridge the gap between intended and experienced HRM, thereby enabling a more accurate assessment of the impact of HR practices on employee attitudes and behaviour (Purcell & Kinnie, 2006; Wright & Nishii, 2007). It also aims at providing a nuanced understanding of the relative impact of people management practices (Wright, Dunford, & Snell, 2001) on employee perceptions of cooperative climate by examining the role of staffing, training and development, and rewards as well as knowledge-work design and immediate management support, two factor which, despite their importance have received little empirical attention (cf. Ramamoorthy & Flood., 2004; Zarraga & Bonache 2005; Cabrera, Collins & Salgado, 2006). More generally, consistent with a relational approach to the HRM knowledge-performance link, it seeks to advance understanding of the breadth and depth of HR systems in a knowledge-intensive organisational context. This article is organised into four sections. The first provides a critical review of the literature from which a set of research questions is derived, accompanied by our proposed model. The second presents the methodology used to test our model. The third presents the results of the study. These are discussed in the fourth section followed by the theoretical implications of our study, its limitations and recommendations for future research.

9.3 HRM and Knowledge-Related Performance

A common goal of recent conceptual and empirical research on the HRM-knowledge-Performance linkage is to explain variation in value creation as a result of coordinating HR with KM strategy. Four distinct approaches are identified in the literature. The first attempts to bridge the gap between HRM and KM by combining theoretical constructs, developed originally in the field of KM, with concepts more familiar within HRM theory. The starting point for building an understanding of explanatory mechanisms is the acknowledgement of the relative importance of different types of knowledge (e.g., explicit, tacit) that are more or less congruent with the strategic priorities of the firm (Hansen et al., 1999). Studies within this perspective reflect a 'best fit' approach to researching HRM-KM linkages (Haesli & Boxall, 2005; Shih &. Chiang, 2005). A second line of work seeks to fill the same gap by Utilising well-established concepts and frameworks from HRM as the basis for developing HR approaches *to* managing knowledge workers. Particular emphasis is placed on the role that high commitment HRM can play in eliciting employee-based capabilities that contribute to the success of KM initiatives. This can be described as the 'best-practice' approach (e.g., Hislop, 2003). As an evolution of the 'best practice' research stream, a third line of work places emphasis on the intermediate role of social relations, culture and climate in the HRM-KM linkage. This can loosely be termed the 'relational' approach (e.g., Zarraga & Bonache, 2005; Cabrera et al., 2006; Collins & Smith, 2006; Kang, Morris, & Snell, 2007). Finally, an emerging body of mainly qualitative studies takes a more critical approach (e.g., Hunter., Beaumont & Lee, 2002; Currie & Kerrin, 2003; Swart & Kinnie, 2003; Willein & Scarbrough, 2006). The relational approach is presented below.

9.3.1 The Relational Approach

Despite a growing consensus that HR systems are the primary means by which firms can manage value-creating social relations (e.g., Lado & Wilson, 1994; Leana & Van Buran 1999; Jackson et al., 2003; Kang et al., 2007), there have been few empirical studies examining whether and how HR practices impact on knowledge flows. A review of the literature identified only a small number of quantitative (Youndt & Snell, 2004; Minbaeva, 2005; Collins & Smith, 2006) and qualitative studies (Hunter et al, 2002; Currie & Kerrin, 2003; Swart & Kinnie, 2003; Willem & Scarbrough, 2006) that have focused explicitly on this area. The first group, which comprises mainly large-scale, survey-based studies (e.g., Youndt & Snell, 2004), examines the relationship between systems of HR practices, social relations and knowledge sharing by seeking to identify "strong situations" (Mischel 1977), such as social capital, that both influence and are influenced by the impact of HR systems on knowledge exchange and, consequently, on organisational performance.

This body of work seeks to explain variation in knowledge sharing effectiveness and performance success as a function of the systemic effects of HR practices on the firm's internal social structure. The second group comprises mainly in-depth, case-based empirical work (e.g., Currie & Kerrin, 2003). While placing equal emphasis on the role of strong situations, it seeks to go a step further by examining the underlying layer of HR processes and how these intertwine with the social context of knowledge sharing. Although the two perspectives are theoretically and analytically different, we believe that they are and should be treated as complementary.

9.3.1.i The systemic perspective

Based on top managers' views of 208 public, single business-unit organisations in the USA, the results of a study by Youndt & Snell (2004) showed that a collaborative based bundle of HR practices were particularly important for enhancing social capital which, in turn, was significantly associated with organisational performance. A closer look at this study indicates, however, three important limitations. First, social capital is operationalised in a rather abstract manner, which makes it impossible to distinguish between its structural, relational, and cognitive dimensions (Nahapiet & Ghoshal, 1998). In fact, Youndt, & Snell (2004) succumb to equating social capital with knowledge sharing. This simplify cation not only hinders understanding of how the distinct dimensions of social capital are shaped differently by HR practices but it also downplays the possibility that, in some cases, knowledge sharing could be a positive spill-over from power and influence relations (portes, 1998; Wlllem & Scarbrough, 2006). A second limitation concerns the poor operationalisation of the HR bundles. The collaborative HR configuration, for example, comprises eight items. This raises questions about the extent to which HR bundles capture adequately the large and diverse array of HR practices required for managing complex social relations. Finally, the study is based on CEOs' views and, therefore, leaves unanswered knowledge workers' perceptions of HR practices and social capital. Based on a sample of 136 high-technology US firms, Collins & Smith (2006) study corrects most of the limitations identified in Youndt & Snell's. (2004) research by developing and testing a more refined model. This model suggests how commitment-based HR practices affect knowledge exchange and organizational performance through social relations. First, commitment-based HR practices are defined here more comprehensively. Second, Collins & Smith (2006) identify organisational social climate as a key mechanism through which commitment-based HR practices affect employee-based capabilities to exchange knowledge. Social Climate is operationalised along three dimensions (i.e., cooperation, trust, shared, language and codes). In this sense, it resembles the relational and cognitive dimensions of social capital (Nahapiet & Ghoshal,1998). Commitment-based HR practices were found to be a strong predictor of all dimensions of social climate. In turn, social climate mediated partially the effect of HR practices on knowledge exchange. In addition, the effect of HR practices on firm performance was mediated not only by knowledge exchange, but also by social climate. Probably the most important contribution of Collins & Smith's (2006) study is that it highlights the crucial role of 'relational social climates' as key mediating mechanisms through which HR systems affect employees' motivation and ability to share knowledge by emphasising that HRM systems are transmitters of core cultural values (Peters, 1978). Notwithstanding its advantages, this study has a number of limitations. First, the composition of commitment based HR practices does not take into consideration aspects, of job design (i.e., reciprocal interdependence autonomy, and variety), which are considered as the defining attributes of knowledge work (Benson & Brown, 2007).

In this sense, the study leaves unanswered how the design of work may condition not only employees' interaction opportunities with others but also their perceptions of social climate and, ultimately, their knowledge sharing, attitudes and behaviour. Second, the study takes an additive approach to testing complementarities between the three sub-facets that comprise the HRM system. In so doing, the possible differential as well as interaction effects

(Ichniowskci, Shaw & Prennushi, 1997) of individual HR practices on social climate are sidestepped. Third, the study focuses only on the HR implications for 'bonding' social capital but provides no guidance on the HR implications for the 'bridging' (Adler & Kwon, 2002) qualities of social relations. Although Collins & Smith (2006) appear to have consciously decided to test their model in firms in rapidly changing industries, the dynamic character of this setting is to a large extent consistent with the entrepreneurial requirements of pursuing exploratory learning. A final and significant limitation of the study is that it downplays the key role that line managers play in influencing employees' experience of HRM.

9.3.1.ii The contextual perspective

Several studies have highlighted the important role of front line management's support in influencing employee knowledge sharing attitudes and behaviour (Hunter et al., 2002; Connelly & Kelloway, 2003;. Zarraga & Bonache, 2005; Cabrera et al, 2006). For example, in Cabrera et al's (2006) study of 372 Spanish employees of a large multinational company, management support emerged as the most important factor affecting knowledge seeking and proving behaviours. Related, a study conducted in five Scottish law firms examining the issue of strategic coordination between HRM and KM showed that the extent to which partners and senior staff were actively involved in knowledge-sharing practice, such as participating systematically in debriefing at the end of projects, sent a strong signal to non-partner staff as to whether knowledge sharing was part of the organisational culture (Hunter et al, 2002). Hunter et al. (2002) conclude that more attention needs to be paid to the management of process upon which informal knowledge sharing depends. Yet, they Argue that, while much of the delivery of HR practices depends on line manage-ment, the HR function does have an important role to play as well. This role, though, is less in the actual delivery than in guiding the professionals, developing consistency of approach and contributing to design (ibid; 18). Achieving balance between the involvement of the HR department and that of line managers in KM practice echoes an important distinction made in the literature between human capital and human process advantage. These are considered as the building blocks of HR advantage (Boxall, 1998). The notion of human process advantage is depicted in Swart & Kinnie's (2003) study of the relationship between HR practices and knowledge sharing in a small software development company in the south-west of England. The key operational processes were distributed across three flat sub-structures (i.e., the committee structure the mentoring structure and the project structure) providing the company with a Unique operational quality which reflected and sustained the organisational routines.

9.3.2 Extending the Relational Approach: Towards Social Resources Management

Consistent with a relational view of competitive advantage (Dyer & Singh, 1998), Kang et al. (2007) have recently introduced a theoretical framework of relational archetypes, namely cooperative and entrepreneurial. These provide the basis for extending the original HR architecture (Lepak & Snell 1999) by identifying two distinct HR configurations pertinent to the management of knowledge flows between core employees and their internal and external partners, respectively. The classic ability-motivation-opportunity (A-M-O) framework, which has guided much 'best practice' research on the HRM-perforrnance relationship (Becker & Huselid, 1998), provides the basis on which Kang et al. (2007) cluster a number of HR practices within each of the two alternative HR configurations. The key difference, however, is that the scope of HR practices expands beyond managing human capital to managing social capital. Essentially, the design of HR configurations is informed by three enabling conditions of knowledge sharing; structural opportunity, cognitive ability, and relational motivation. These conditions are, in turn, reflected in three HR practice areas: (i) work design structures (e.g., job variety, autonomy, interdependence), (ii) incentive structures (e.g., pay, performance appraisal, employment security), and (iii) skill development (e.g., staffing, training, mentoring). By identifying two relational archetypes and the ways through which they are supported by two distinct configurations of HR practices, Kang et al. (2007) contribute significantly to a better understanding of the HRM Knowledge-performance linkage by placing explicit focus on the mediating role of value-creating social relations. There model reframes the problem the modem HR function faces as it strives to balance between efficiency and flexibility. (Rousseau & Arthur, 1999). The two relational archetypes and their associated HR configurations are, however, only theoretically derived and, therefore, deserve empirical investigation. There are still a number of important issues that remain unresolved.

The relational archetypes, as the term implies, are ideal types. However, in practice it may be difficult to draw a clear line between cooperative and entrepreneurial social relations. As Evans & Davis (2005: 772) note: 'dynamic environments appear to be more the norm than the exception for organizations, limiting the applicability of the boundary condition'. Empirical evidence also suggests that organisations are likely to implement hybrid HR systems, particularly with respect to their core knowledge employees (Lepak & Snell, 2002). A key question, therefore, concerns the extent to which HR practices comprising seemingly coherent HR bundles send contradictory messages to core knowledge employees as to which types of social relations are most valued. As Kang et al. (2007) suggest knowledge employees may differ from organisational strategists in their views of which type of social relations are most valued and rewarded. This brings to the forefront not only the complicated issue of demarcating employment modes and specifying which employee relations constitute the core competence of the firm, but also the importance of focusing on how employees experience HR practices. In addition, while recent empirical evidence demonstrates the additive effects of commitment-based HR practices on cooperative social climate (Collins & Smith, 2006), the literature lacks a systematic study of the individual effects of HR practices on employee perceptions of that climate. It is, therefore, important to disaggregate the HR bundles and examine the influence of each HR practice on employee perceptions of organisational social Climates favourable to knowledge sharing.

9.3.2.i Social climate considerations

A closer look at Kang et al's (2007)'model suggests that the two relational archetypes reflect two different kinds of organisational climate. Specifically, in the cooperative archetype, which is underpinned by a collectivist culture, social relations are based on strong norms of cooperation and reciprocity, mutual trust and identification. On the other hand in the entrepreneurial archetype, which reflects a somewhat more individualistic or ego-centric culture, social relations can be viewed more as an asset that 'inheres in a focal actor's external network that give the actor advantages in his or her competitive rivalries' (Xiao & Tsui, 2007: 3). What is therefore missing from Kang et al's (2007) Conceptual framework is an explicit emphasis on the social context by which HR systems are shaped. The term social context 'embodies the very essence of organizational science and, as such, serves as an effective mechanism through which to more precisely articulate how HR systems relate to organization effectiveness' (Ferris, et al, 1998: 237, 239). A social context approach to HRM encompasses culture, climate and, more broadly, social and political processes as essential features of work environments that contribute to organizational effectiveness. Accordingly, the core values, assumptions, beliefs, and political issues that comprise the culture of the organisation shape the design and implementation of HR policies and practices. For example, HR systems can be characterised by a stronger concern for employee welfare and a weaker focus on task performance expectations. (Von Glinow, 1985). Performance evaluations in 'caring' HR systems focus less on criteria such as in-role performance, and more on criteria of contextual performance such as teamwork, cooperation and, cultural fit (VonKrogh, 1998; Zarraga & Bonache, 2005). According to social context theory, HR practices shape employee attitudes and behaviour mainly through their impact on employees' interpretations of the organisational climate. This refers to the 'more temporary and changeable interpretation of an environnient by participants operating within that context' (Ferris et al., 1998: 243). A core premise of the social context approach is that the extent to which HR practices affect one or more of the dimensions of the organisational climate depends on the extent to which these practices are internally consistent and reflective of the wider organisational culture. While the HRM-culture linkage is usually present in the formulation of HR policies, the strength of that linkage may be weakened during the implementation of HR practices as this is reflected in the impact of HR practices on organizational climate. This can result from 'errors of commission' whereby multiple stakeholders, particularly line managers, may use the HR system politically to satisfy agendas other than operational effectiveness (ibid.).

9.3.2.ii Research questions and proposed model

Underpinned by a social context approach to the: HRM–knowledge-performance linkage, the aim of our study is to understand the effects of HR practices, as experienced by employees, on their perceptions of organisational social climate of teamwork and cooperation and, by extension, on knowledge sharing attitudes and behaviour. While recent empirical studies suggest that commitment-based HR systems have a positive impact on teamwork and cooperation climate (Collins & Smith, 2006), the possibility that each of the HR practices comprising the HR system may exert differential influence on that climate remains largely unexplored. Furthermore, despite theoretical and empirical support for the catalytic role, that line managers play in the successful delivery of HR practices (Arthur & Boyles, 2007; Purcell & Hutchinson, 2007); very few studies have examined the possibility that the effect of immediate management support for knowledge sharing on employee perceptions of a social climate of teamwork and cooperation may be similar to or even more important than the effect of HR practices (e.g., Cabrera et al, 2006). This article focuses on two key questions that remain unanswered: (1) what are the individual effects of employee perceptions of HR practices on their perceptions of a cooperative social climate conducive to knowledge sharing attitudes and behaviour? (2) What is the relative importance of employee perceptions of line management support for knowledge sharing on that climate? Based on these questions, we developed a model which is illustrated in Figure 1.

Taken together, the aforementioned questions address; (i) the issue of differential effects of HR practices on employee perceptions of a cooperative social climate, (ii) the issue of expanding the scope of HR systems to include the role of job design and line management support as key antecedents of that climate and (iii) the issue of conflicting messages that hybrid HR systems may send to knowledge workers with regard to which behaviours are encouraged and valued. Answers to these issues will help shed valuable light on the HRM knowledge performance link by identifying; (i) the possibility that various HR practices may impact to varying degrees on the creation of a cooperative social climate conducive to knowledge- sharing, (ii) the potentially significant role that line managers play not only in fostering such a climate but also in mediating the effect of HR practices on that climate, and (iii) the possibility that employees ascribe the role of 'relationship builder' mainly to line management (Lengnick-Hall & Lengnick Hall, 2003).

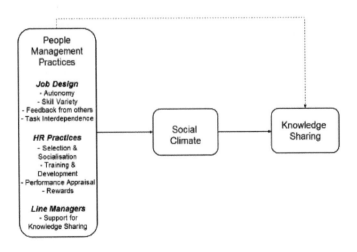

Figure 1 Proposed Model

9.4 Methods

9.4.1 Setting and Sample

The study was conducted in units of three organisations located in Ireland; the management consultancy unit of a professional services firm (hereafter, Consult Co); the network engineering unit of a telecommunications company (here after, Teleco); and the headquarter offices of a semi-state business development agency (hereafter, State Co). An online qestionnnaire survey was conducted with employees from the three organisations between February and July 2005. A total of 563 surveys were sent to the three organisations, 135 of which were completed successfully and submitted on-line - 43 from ConsultCo, 58 from TeleCo and 37 from StateCo. The overall response rate was 24.5% ranging from 17% for StateCo, to 23% and 48% for TeleCo and Consult Co, respectively. In addition, qualitative data were collected by conducting six semi-structured interviews with the senior HR managers and KMproject managers within the organizations. The final sample consisted of full-time, core employees engaged in knowledge-intensive work (i.e., management consulting, IT engineering, strategic planning) organised in a project-based fashion. Project-based work is viewed as increasingly important for the successful coordination of the complex, interdependent and non- routine of knowledge-intensive work activities (Turner 1999; Benson & Brown, 2007). The sample was gender balanced (49.5% women) with an average age of 35 years (range 23–60 years). The majority of the sample (95%) had a third-level educational qualification either at postgraduate level (52%), undergraduate level (34%) or diploma level (9%). Almost half of the respondents were employed in management-level positions (51%), while 49% described their jobs as professional (31%), technical (10%), and support (8%). The average organisational and positional tenure of the sample was 8.5 and 2.5 years respectively, with an average industry experience of 12 years.

9.4.2 Measures

Perceptual measures were used to gauge employee-level experiences of job design HR practices, management support for knowledge sharing and organisational social climate using multi-item constructs, rated on seven-point likert-type scales. With the exception of HR practices, all constructs were adopted from pre-existing scales found in the literature. All items were-factor-analysed using maximum likelihood with promax rotation to examine the psychometric properties of the measures, focusing on-dimensionality, and reliability. The derived measures achieved satisfactory internal consistency levels (Cronbach alphas >.70).

9.4.2.i Job Design

Measures for autonomy, skill variety, and feedback from others were adapted from Idaszak & Drasgow's (1987) revised version of Hackman & Oldlianvs (1975, 1980) job diagnostic survey (JDS). The revised version corrects the weaknesses of the original JDS by replacing reverse coded items with positively worded ones. Pearce & Gregersnen's (1991) scale was used to measure reciprocal task interdependence.[1]

9.4.2.ii HR Practices

Conceptual and empirical studies examining the links between HRM, social relations/social climate, and KM provided the basis for developing measures of employee perceptions of relational HR practices (e.g., Leana & Van Buren, 1999; Zarraga & Bonaclie, 2005; Kang et. al, 2007). 18 original items were devised around four HR practice clusters: selection and socialisation, training and development, performance appraisal, and rewards. Each item asked respondants to indicate on a seven-point Likert-type scale the extent to which they had experienced a specific HR practice.[2]

9.4.2.iii Management Support for Knowledge Sharing

Connelly & Kelloway's (2003) six-item measure was used to assess employee perceptions of management support for knowledge sharing. Three items focus on immediate manager support for eliciting employees' knowledge sharing behaviours, while the remaining three items focus on more formal, systemic aspects of organisational support for knowledge sharing. All items

---

[1]The result of factor analysis which are available upon request, produced three instead of four factors as it \would normally be expected. 'While the 3 items making up the 'feedback from others' scale, and the 5 items comprising the 'task interdependence, scale loaded strongly into the right constructs, the 3 items corresponding to the job autonomy scale loaded into the same factor as the 3 items comprising the skill variety scale. Given the lack of a clear factor structure with regard to job autonomy and skill variety, it was decided to exclude both measures from further statistical analysis.

[2]The results of factor analysis indicated a clear structure for all items with the exception of performance appraisal (two items), which, as a result, was excluded from further statistical analysis. The 18 items and their wording are provided in the Appendix.

loaded on a single factor providing support for the discriminant validity of the measure.

9.4.2.iv Team work and cooperation climate

Valle & Vitte's (2001) three-item construct was used to assess individual perceptions of the importance of cooperation and team orientation within the organisation. This measure exhibited good discriminant validity as all items loaded on a single factor. Finally, based on the results of one-way between groups analysis of variance (ANOVA), there were no significant differences found across the three organisations.

9.4.2.v Control variables

A set of demographic variables were also included in the survey. Respondents were asked to indicate their age, gender, educational qualification, job type, organisational as well as positional tenure and industry work experience.

9.5 Results

Table 1 presents discriptive and skewness statistics, internal reliabilities and inter -correlations among the variables of interest. All skewness statistics were found to be less than 1.0, which suggests that the variables were relatively normally distributed (Miles & Shelvin, 2001).

Table 2 provides a summary of the results of regression analyses regarding the partial

And overall effects of independent variables on employee perceptions of teamwork and cooperation climate.

As shown in table 2, the job design variables explained almost a quarter of the variance in the outcome variable,[3] with both reciprocal task interdependence and feedback form others emerging as significantly positive predictors of teamwork and cooperation climate. The HR practices explained 34% of the variance in the outcome variable. However, only selection and socialisation, and type of training and development were significantly associated with teamwork and cooperation climate.

Finally, management support for knowledge sharing accounted for 29% of the variance in the outcome variable. Furthormore, it not only remained a

---

[3]The Sole effect of control variables on the outcome Variable was found to be negligible

Table 1   HR and Associated Variables: Means, Standard Deviations, Skewness, Correlations

Variables	Means (SD)	Skew	1	2	3	4	5	6
1. Task Interdependence	6.01 (.81)	.95	(.82)					
2. Job Feedback	4.56 (1.30)	−.45	.12	(.81)				
3. Selection and Socialisation	4.46 (1.17)	.34	.21*	.37**	(.70)			
4. Quantity of Training and Development	3.91 (1.58)	.06	.08	.22*	.20*	(.84)		
5. Type of Training and Development	4.15 (1.05)	−.33	.24**	.36**	.40*	.43**	(.68)	
6. Rewards Mix	3.33 (1.32)	.01	−.05	.50**	.38**	.25**	.40**	(.82)
7. Rewards Competitiveness	3.48 (1.31)	.10	−.06	.08	.32**	.27**	.28**	.35*
8. Rewards Equity	4.01 (1.49)	.03	−.05	−.04	−.06	−.11	.09	.04
9. Support for Knowledge Sharing	4.23 (1.09)	−.13	.19*	.40**	.33**	.22**	.48**	.25*
10. Teamwork and Cooperation Climate	4.64 (1.25)	−.55	.27**	.41**	.46**	.25**	.39**	.30*

N=135; Two-tailed tests, **p<.01; *p<.05; Internal reliabilities are shown along the diagonal in parentheses.

Table 2 Regression Results

Independent Variables	B^a	B^b	R^c	R^d
				$\cdot42^{***}$
Demographics (not shown)				
Job Design			$\cdot26^{**}$	
Task Interdependence	$\cdot34^{**}$	$\cdot21$		
Feedback from Others	$\cdot35^{***}$	$\cdot14$		
HR Practices			$\cdot34^{***}$	
Selection and Socialisation	$\cdot47^{***}$	$\cdot29^{*}$		
Training and Development (Quantity)	$\cdot05$	$\cdot04$		
Training and Development (Type)	$\cdot22^{*}$	$\cdot03$		
Rewards (Mix)	$\cdot06$	$\cdot03$		
Rewards (Competitiveness)	$-\cdot04$	$\cdot04$		
Rewards (Equity)	$-\cdot10$	$-\cdot09$		
KM Practices			$\cdot29^{***}$	
Management Support for Knowledge Sharing	$\cdot58^{***}$	$\cdot33^{**}$		

Notes:

[a] Standardised beta weights controlling for demographic variables and/or other variables within the same set.

[b] Standardised beta weights controlling for demographic variables and all other variables.

[c] R Square for all variables within a set controlling for demographic variables.

[d] R Square for all variables within a set controlling for demographic variables and all other sets.

***$P<001$; **$P<01$; *$P<05$.

significant and positive predictor of teamwork and cooperation climate when controlling for the rest of the variables, but it also suppressed the prior positive effect of job design and type of training and development, yet not of selection and socialization. Overall, job design, HR practices, and management support for knowledge sharing explained 42% of the variance in teamwork and co-operation climate, which is indicative of the strong explanatory power of our proposed model. The sole effect of control variables on the outcome variable was found to be negligible social climates. Importantly, it highlights that not all of the HR practices that comprise an HR system are equally important in terms of their effects on employee perceptions of teamwork and cooperative climate.

The results indicate that, on the one hand, selective hiring and intensive socialisation, and relational-oriented training and development send strong signals to employees regarding the importance of teamwork and cooperative spirit for governing work interactions. However, on the other hand, the relative weight of these practices on employee perceptions of teamwork and cooperation weakened and in the case of training and development disappeared, in the presence of high reciprocal task interdependence and of an effective multi rater job feedback system. Taken together, the findings suggest that, in essence, job and team design structures can be seen as alternative methods for evoking prosocial behaviours, such as knowledge shairing, through producing strong perceptions of a social a social climate that values and encourages a cooperative spirit among employees. While several scholars suggest that the best means to support knowledge sharing in organisations is to hire smart people and let them talk to one another (Davenport & Pusak 1998), we go a step further and add to the above suggestion by concluding that line managers play a vital role in encouraging employees to "talk to one another". Our findings confirm the need for extending the notion of the HR system to include the catalytic role of line managers in 'influencing perceptions not only of HR practices but of work climate' (Purcell & Hutchinson, 2007:5). In this regard, our research is one of the first efforts to add to this extra dimension to the HRM-knowledge-performance link, thereby providing substantive support for the claim that people management is the combination of leadership behaviour, HR practices and organisational climate' (ibid: 17)

9.5.1 Limitations and Directions for Future Research

The results presented in this article are however limited, in that they shed light only on the role of employee perceptions of HR practices on teamwork and cooperation climate, but without observing how that climate is associated with knowledge sharing attitudes and behavior. Additional research is required to estabish further that link. A second limitation is related to the operationalisation of HR practices. While these were loosely clustered around ability, motivation and, opportunity, specific measures for ability, motivation, and opportunity are required in order to determine the exact pathways through which HR practices affect teamwork and cooperation climate and knowledge sharing. (Siemsen, Roth, & Balasubramanian, 2008). The employment of measures for purposeful, actionable knowledge sharing (Cross & Sproull, 2004) would add significantly to a deeper understanding of the HRM knowledge flows linkage. Moreover, while our focus was place on the role of HR practices on developing social relations, further work is required to shed light on the complementarities as well potential conflicts with respect to the management of human and social capital. For example, future studies could examine the issue of complexity of knowledge governance mechanisms and its implications for the design of HR systems congruent, with the management of cooperative as well as entreprenerial social relations (Truss, 2001; Foss, 2007; Kang et al, 2007).

Finally, from a methodological standpoint, the results are limited in their generalisability because of the small sample size. An additional limitation is related to common method bias due to the use of self-report measures of both independent and dependent variables obtained from a single source. Although the results of Hartman's one-factor test indicated the absence of a single-factor, common method bias may not have been completely removed in the study.

9.6 Conclusion

This study contributes to a better understanding of the breadth and depth of HR systems in a knowledge-intensive organisational context. In terms of breadth, it suggests that the role of line managers lies at the heart of the HRM-KM relationship since it is mainly line managers' behaviour that serves as a core basis on which employees develop shared understandings of a social climate where teamwork and cooperation are desired and valued by the organisation. In addition, it shifts attention to the fundamental role of the design of knowledge work as a building block of employee perceptions of that climate. Finally in terms of depth, the study suggests that the effective management of social relations may require a process-based HR approach that goes beyond explicit motivation mechanisms, such as pay incentives for sharing knowledge, and directs attention to core structural aspects of knowledge work as well as to softer incentives for supporting prosocial behaviours and value-creating social relations.

References

1 Argote, L, & Ingrain, P: 2000. Knowledge transfer: A basis for competitive advantage in firms. Qrganisational Behavior and Human Decision Process, 82:150–169.
2 Arthur, JB, & Boyles, T, 2007. Validating the human resource system structure: A Levels-based strategic HRM approach. Human Resource Management Review, 17:77–92.
3 Becker, B., E., & Huselid., M. A. 1998. High, performance work systems and firm performance: A systhesis of research and managerial implications, Research in Personal and Human Resource Management, 16:53–101.

4 Benson, J., & Brown, M. 2007. Knowledge workers: What keeps them committed; what turns them away. Work, Employment & Society 21(1); 121–141.

5 Boxall, P. 1998. Achieving competitive advantage through human resource strategy: Towards a theory of industry dynamics. Human Resource ManagementReview 8(3): 256–288.

6 Boxall, P., & Purcell J, 2003. Strategy and human resource management, Basingstoke: Palgrave Macmillan.

7 Brass, D. J. 1995. A social network perspective on human resource management. In G. R. Ferris (Ed.), Research-in Personnel and Human Resource Management 39–79, Greenwich: JAI Press.

8 Brass, D. J., & Labianca, G. 1995. Social capital, social liabilities and social resources management. In R. T. A. J. Leenders and S. M. Gabbay (Eds,), Corporate social capital and liability: 323–341. Boston: Kluwer Academic.

9 Cabrera, A., Collins, W., C., & Salgado, J. F. 2006. Determinants of individual engagement in knowledge sharing. International Journal of Human Resource Managemmt, 17(2): 254–264.

10 Collins, C. J., & Smith, K, G, 2006. Knowledge exchange and combination: The role of human resource practices in the performance of high-technology firms Academy of Management Journal, 49(3): 544–560.

11 Connelly, C. E., & Kelloway, E. K. 2003. Predictors of employees' perceptions of Knowledge sharing cultures. Leadership & Organization DevelopmentJournal 24(5): 294–301.

12 Conway, J. M. 1999. Distinguishing contextual performance from task performance managerial jobs. Journal of Applied Psychology 84: 3–13.

13 Cross, R., & Sproull, L, 2004. More than an answer: Information relationships for, Actionable knowledge. Organization Science, 15(4): 446–462.

14 Currie, G. & Kerrin., M. 2003. Human resource management and knowledge management: Enhancing knowledge sharing in a pharmaceutical company International Journal of Human Resource Management, 14(6): 1027–1045.

15 Davenport T. H., & Prusak, L, 1998. Working Knowledge: How organizations manage what they know. Boston: Harvard Business School.

16 Evans-W, R., & Davis, W, D. 2005. High-performance work Systems and Organizational performance: The mediating role of internal social structure. Journal of Management, 31(5): 758–775.

17 Ferris, G, R., Arthur, M. M., Berkson, H. M., Kaplan, D. M., Harrel- Cook, G., & Frink, D. D. 1998. Toward, a social context theory of the human

resource management- organizational effectiveness relationship, Human Resource Management Review. 8(3): 235–264.

18 Freeman R. B. & Weitzman, M. L, 1987. Bonuses and employment in Japan. Journal of the Japanese and International Economies 1:168–194.

19 Hackman, J. R. 1987. The design of work teams. In J. Lorsch (Ed,), Handbook of organizational behavior. Englewood Cliffs, NJ: Prentice-Hall.

20 Hackman, J. R., & Oldham G. R. 1975. Development of the job diagnostic survey. Jounal of Applied Psychology, 60(2): 159–170.

21 Hackman J. R., & Oldham, G. R. 1980. Work redesign, Reading MA: Addison-Wesley.

22 Haesli, A., & Boxall, P. 2005. When knowledge management meets HR strategy: An exploration of personalisation- retention and codification-recruitment. Configurations. International Journal of Human Resource Management 16(11): 1955–1975.

23 Hansen, M, T., Nohria, N., & Tierney, T, 1999. What's your strategy for managing Knowledge? Harward Business Review, March-April: 106–116.

24 Hargadon, A., & Sutton, R, 1997. Technology brokering and innovation in a product development firm, Administrative Science Quaterly, 42: 716–749.

25 Hislop, D; 2003. Linking human resource management and knowledge management Via commitment: A review and research agenda. Employee Retention, 25(2):182–202.

26 Hunter, L., Beaumont, P., & Lee, M. 2002. Knowledge management practice in Scottish law firms. Human Resource Management Jounrnal 12(2): 4–21.

27 Ichniowski, C, Shaw, K., & Prennushi, G. 1997. The effects of human resource management practices on productivity: A study of steel finishing lines. The American Economic Review, 87(3): 291–313.

28 Idaszak, J. R., & Drasgow, F, 1987. A revision of the job diagnostic survey: Elimination of a measurement artifact. Journal of Applied Psychology 72(1):69–74.

29 Jackson, S. E., Hitt, M. A., & DeNisi, A. 2003. Managing human resources for knowledge- based competition. In S. E. Jackson, A. DeNisi and M. A. Hitt (Eds.), Managing knowledge for sustained competitive advantage: 399–428. San Francisco, CA: Jossey-Bass.

30 Jackson, S. E., Chuang, C.-H., Harden, E. E, & Jiang, Y. 2006. Toward developing human resource management systems for knowledge-intensive teamwork. Research in Personal and Human Resource Management, 25: 27–70.

31 Kang, S.-C., Morris., S. S., & Snell, S. A. 2007. Relational archetypes, organizational learning, and value creation; Extending the human resource architecture. Acedemy of Management Review, 32(1); 236–256.

32 Lado, A. A., Wilson, M. C. 1994. Human resource systems and sustained competitive advantage: A competency-based perspective. Academy of Management Review, 19(4): 699–727.

33 Leana, C. R., & Van Buren III, H. J. 1999. Organizational social capital and employment practices. Academy of Management Review, 24: 538–555.

34 Lengnick-Hall, M. L., & Lengnick-HalL, C. A. 2003. Human resource management in the knowledge economy: New challenges, new roles, new capabilities san Francisco: Berret-Koehler.

35 Lepak, D. P., & Snell, S. A. 1999. The human resource architecture; Toward a theory of human capital allocation and development. Academy of Management Review, 24: 31–48.

36 Lepak, D. P., & Snell, S. A. 2002. Examining the human resource architecture: The relationship among human capital, employment, and human resource configurations . Journal of Management, 28(4): 517–543.

37 Minbaeva, D. 2005. HRM, practices and knowledge transfer. Personnel Review, 35(1):125–144

38 Mischell, W. 1977. The interaction of person and situation. In D. Magnusson and N. S. Endler (Eds.), Personality at the cross road: Current issues in interactional psychology: 333–352. Hillsdale NJ: Erlbaum.

39 Nahapiet, J., & Ghoshal, S. 1998. Social capital, intellectual capital, and the organizational advantage. Academy of Management Review 23(2): 242–266.

40 Nahapiet, J., Gratton, L., & Rocha, H. O. 2005. Knowledge and relationship. When cooperation is the norm. Europion Management Review, 2:3–14.

41 Nonaka, I., & Takeuchi, H. 1995. The knowledge-creating company: New York: Oxford University Press.

42 Pearce, J. L. & Gregersen, H. B. 1991. Task interdependence and extra-role behavior: A test of the mediating effects of felt responsibility. Journal of Applied Psychology, 76(6): 838–844.

43 Peters, T. J. 1978. Symbols, patterns, and settings: An optimistic case for getting things done. Organizational Dynamics,1:3–23.

44 Portes, A. 1998. Social capital: Its origins and applications in modem sociology. Annual Review of Sociology, 24, 1–24.

45 Purcell, J., & Hutchison, S. 2007. Front-line managers as agents in the HRM performance causal chain: Theory, analysis and evidence. Human Resource Management Journal, 17(1): 3–20

46 Purcell, J., & Kinnie, N.2006. HRM and business performance. In P Boxall, J.Purcell and P. M.Wright (Eds.), The Oxford handbook of human resource management Oxford: Oxford. University Press.

47 Ramamoorthy, N. & Flood, P. C, 2004. Individualism/collectivism, perceived task interdepen-dence and teamwork attitudes among Irish blue-collar employees: a test of the main and moderating effects, Human Relations, 57(3): 347–366.

48 Rousseau, D., & Arthur, M. B. 1999. The boundaryless human resource function: Building agency and community in the new economic era. Organizational Dynamics, Spring: 7–18.

49 Shih, H.-A., & Chiang, Y.-H: 2005. Strategy alignment between HRM. KM, and corporate development. International Journal of Manpower, 26(6): 582–603.

50 Storey, J. & Quintas, P. 2001. Knowledge management and HRM. In J, Storey (Ed.), Human resource management: A critical text 339–363. London: Thomson Learning.

51 Svetlik, I., & Stavrou-Costea, E. 2007. Introduction: Connecting human resources management and knowledge management. International Journal of Manpower, 28(3/4):197–206.

52 Swart, J., & Kinnie, N. 2003. Sharing knowledge in knowledge-intensive firms. Human Resource Managemant journal 13(2): 60–75.

53 Truss, C. 2001. Complexities and. controversies in linking HRM with, organizational outcomes. Journal of Management Studies, 38(8): 1121–1149.

54 Turner, J. R. 1999. The handbook of projectbased management: Improving the process for achieving strategic objectives (2nd Eel), London: McGraw-Hill.

55 Valle, M., & Witt, L. A, 2001. The moderating effect of teamwork perceptions on theOrganizati-onal politics-job satisfaction relationship. The Journal of Social Psychology, 141(3):379–388.

56 Von Glinow, M. A. 1985. Reward strategies for attracting, evaluating and retaining professionals. Human Resource-Management 24:1191–1206.

57 Von Krogh, G. 2003. Knowledge sharing and the communal resource. In M, Easterby-Smith and M. A. Lyles (Eds.), The Blackwell handbook of organizational learning and knowledge management 372–392. Oxford: Blackwell.

58 Vroom, V. H. 1964. Work and motivation. New York: Wiley.

59 Wageman, R., & Baker, G.1997. Incentives and cooperation: The Joint effects of task and reward interdependence on group performance. Journal of Organizational Behavior, 18(2): 139–159.

60 Willem, A., & Scarbrough, H. 2006. Social capital and political bias in knowledge sharing: An exploratory study. Human Relation, 59(10): 1343–1370.

61 Wright, R. M., Dunford, B. B, & Snell, S. A; 2001. Human resources and the resource based view of the firm. Journal of Management 27: 701–721.

62 Wright, P., Nishi, L, 2007. Strategic HRM and organizational behavior: Integrating multiple levels of analysis. Working Paper 07–03, CAHRS Working Paper Series, Cornell University, NY.

63 Xiao, Z., & Tsui, A, S. 2007. When brokers may not work: The cultural contingency of social capital in Chinese high-tech firms. Administrative Science Quarterly, 52:1–31.

64 Youndt, M. A., & Snell, S. A, 2004. Human resource configurations, intellectual capital, and organizational performance, Journal of Managerial Issues, 16(3): 337–360.

65 Zarraga, C. & Bonache, J. 2005. The impact of team atmosphere on knowledge outcomes in self-managed teams. Qrganizational Studies 26(5): 661–681.

66 Zupan, N. & Kase, R, 2007. The role of HR actors in knowledge networks. International Journal of Manpower, 28(3/4): 243–259.

10

SMS'S Innovation and Human Resource Management

Professor Marius Dan Dalota, Ph.D.

Romanian-American University,
1B, Expozitiei Avenue, Sector 1, Bucharest, Romania

10.1 Abstract

This paper aims at analysing the relationship between innovation and human resource management (HRM), attempting to establish whether innovation determines the firm's human resource management or, conversely, human resource management influences the innovation level of the company. Based on this review, some research hypotheses are formulated. Article's findings results provide evidence that, in order to affect employee behaviour, the firms must develop a bundle of internally consistent HRM practices.

Keywords: Innovation; Small to medium-sized enterprises; Human resource management; Classification:

10.2 Introduction

Innovation and human resources management play an increasingly important role in sustaining "leading edge" competitiveness for organisations in times of rapid change and increased competition. "Discontinuous change requires discontinuous thinking. If the new way of things is going to be different from the old, not just an improvement on it, then we need to look at everything in a new way". The continuous hegemony of innovation and creativity arises from organizations recognizing that correctly harnessed creativity can offer companies a competitive advantage (Porter, 1980). The analysis of the strategies of the top companies of the future, the structural flexibility and innovative power were listed among the top drivers of future success. Today, firms are facing a competitive and continuously changing situation. In this context the performance, and even the survival, of firms depend more than ever on their ability to achieve a solid and competitive position and on their flexibility, adaptability and responsiveness. Therefore, it is hardly surprising that there is growing interest in innovation as a strategy that allows the firm to improve its flexibility, competitive position and performance. A company will create new products for a variety of reasons, but usually in an attempt to increase profits. The most profitable new products will be those that meet the customer needs more effectively than competitors' products, and are therefore preferred by more customers. Companies need to identify those needs, and then generate ideas and solutions to address them. Many articles on innovation and creativity begin with a general statement that companies must innovate or they will die. While this is generally true, any company that is inefficient in vetting and implementing new product ideas or a company that continually introduces the wrong products will consume its own resources and will also fail.

In the management field, a host of variables has been identified as influencing organizational innovation:

- having a vision of what the organization wishes to be,
- searching for opportunities, experience and technological potential
- following market orientation,
- market evolution and segmentation and the promotion and management of creative resources.

Human factors and, in particular, human resource management, are today considered key elements of successful innovation, since the human element is involved in the whole innovation process. Specialists state that there are no good technologies or good innovations without competent people who can adequately use them and get benefit from them. At the same time, no competent people can be available if there is not, first, a business project defining the role that technology and innovation must play and creating the necessary and sufficient conditions for channelling aptitudes, capacities and attitudes of the individuals towards the established direction.

Considering that HRM determines and modifies, to a large extent, these aptitudes, capacities and attitudes, it seems clear that it becomes a crucial element in the development of innovation activities. Human resource management has been, up to now, scarcely treated in studies on innovation in the firm. Although there have been some empirical studies in recent years, their conclusions are heterogeneous and most of them have focused on U.S. firms.

10.3 Dimensions of SMS's Growth Through Innovation

Three main sources of growth can be determined:

a) *Technological improvement* - It is well known that processes and technology improvements can contribute to meeting quality and process - performance, objectives.

b) *An increase in the quantity of capital* - Very often, technology is deeply linked to investment because it is embodied in new machinery and better equipment.

c) *An increase in the number of workers, their skills and educational levels.* Industry growth depends on several internal and external factors, such as physical assets, technologies all along the chain value, human resources in general and qualification levels in particular and also organizational capabilities. In general, the firms are more likely to reap profits and social benefits when they are in high-growth industries.

SMEs can increase their activities and businesses in some ways and grow in some dimensions. The following dimensions can be identified:

a) *Raise the level of integration of the technologies* - The management of technologies and the exploitation of all their potential is strictly linked to the possibility of integrate their synergies.

b) *Intensify innovative technology processes* - This direction of innovation is a decisive contribution for the modernization of businesses and the implementation of competitive strategies.

c) *Increase the number of markets where the company operates*- Internationalization and globalization are direct consequences of this decision.

d) *Increase businesses portfolios* — The company that today is involved in a given industry can tomorrow widen its investment to other industries.

e) *Increase the number of operational uses of technologies* – Many technologies can have applications in operations of a different nature. To position strongly for future growth in the global marketplace, an organization has to make some effort to increase its investments in R&D and to focus on the implementation of advanced production innovations and practices. The growth of an organization, the technologies that are being used along all its activities, and business strategies that have been formulated are strictly related. Even organizational culture deals with technologies and growth.

Technological progress driven by a decision to enhance productivity and profitability often fosters growth. The competitive success of most enterprises is strongly related to decisions such as:

- producing products and services according to high quality standards;
- quantifying production in the correct manner;
- anticipating and responding to changing consumer needs;
- reducing production costs in order to enhance profitability.

The growth effort has to include:

- New technologies for manufacturing with ecological safety.
- Designing and modelling of secure facilities.
- Adopting zero-waste procedures in manufacturing and processing,
- Upgrading of existing installations.
- Developing new organizational tools and methodologies,
- Reducing resource consumption in order to reach competitive production costs.

The success of SMEs depends on:

- using advanced technologies in an integrated manner,
- being aware of changing clients' needs, producing quality goods and services
- enhancing profits by reducing costs,
- reaching new markets within a competitive perspective,
- wide-open mentality

Many SME's are not able to envisage growth as a competitive need and this difficult mentality and/or reluctance should be understood.

The identification of innovative improvements that could enhance organizations' movements for growth is a decisive process to reach growth objectives. Innovation in production, distribution, and communication processes serve as a vital source of productivity growth and other competitive advantages. The success of most management innovation processes is also a function of competitive efforts.

The managerial decision regarding obtaining growth results has to take into account what is needed to reach a rapid modification in the professional qualification levels of workers and managers. It is indispensable that a strict and dedicated cooperation exists among governmental entities, industries and educational sectors. When an entrepreneur does not have experience and technical knowledge in the financial domain he may have a distorted perception of the reality, because an increase in sales does not necessarily correspond to an increase in profitability and, therefore, does not open the possibility of self-financing. It is well known that some entrepreneurs prefer self-financing because it provides them with more control. It is required to create new higher education models in the domain of entrepreneurship. We agree that a new higher education models will require the commitment of governments, universities, and associations.

10.4 The Relationship between Innovation and HRM

There are no best HR practices, because in order to be effective, HR practices must be consistent with other aspects of the organisation, specifically its strategy. The most suitable HRM practices for firms trying to develop a competitive advantage based on innovation will be different to those practices suitable for firms seeking other kind of competitive advantage.

Some studies have focused on some isolated HRM practices while others have focused on the HR system adopted by the firms. The underlying assumption of these studies is that the impact on organizational performance of sets or "bundles" of interrelated HR practices can be greater that the cumulative impact of all the individual practices comprising the bundle. Most of the above mentioned studies are based on the models proposed by well-known specialists (Miles and Snow and Schuler and Jackson). Their model proposes the development of a market-type HRM system for those firms defined as prospectors, firms characterized by the search for new products and markets, which are, therefore, the innovators. They argue that it is very difficult for the firm to provide the necessary abilities for a new market or product from inside the company and, more importantly, to provide them quickly. Therefore, these authors recommend searching outside for these abilities whenever the organization needs them, i.e. developing a market-type HRM system.

Other specialists (Schuler and Jackson) establish a connection between HRM practices and three types of strategy: costs, quality and innovation, defined from Porter's (1980) classification of competitive strategies. Their model starts by analysing the employee's behaviour required by each kind of strategy; subsequently they propose HRM practices for the development of these behaviours. In their view, when contemplating an innovation strategy, the firm needs creative employees who are flexible and tolerant of uncertainty and ambiguity; people who are able to take risks and assume responsibilities, very skilful, able to work in a cooperative and interdependent way and with a long-term orientation. The HRM practices that those systems include are listed in Table 1. We can say that, between specialists, there is agreement, first,

Table 1   HRM Practices for Innovation

	Miles and Snow's Model	Schuler and Jackson's Model
Recruitment and selection / *Training*	Emphais: "buy" Hiring almost exclusively from Outside the organization Selection may involve pre-employment psychological testing very little employment security given Little if any sociali-sation taking place within the organisation Skill identification and acquisition	External source of recruitment Technical and research competencies High employment security
Development and Internal Career Opportunities / Performance Appraisal / Compensation	Limited training programme very little use of internal career ladders Results-oriented procedure Identification of Staffing needs Division/corporate performance evaluations Cross-sectional compar-isons Oriented towards performance External com-petitiveness Total compensation heavily oriented towards incentives and driven by recruitment needs	Broad application Employees are responsible for learning Jobs that allow employees to develop skills that can be used in other positions in the firm Broad career paths Mandatory competency growth Process and results criteria Performance appraisals that are more likely to reflect longer-term and group -based achievements Many incentives Internal equity Low pay rates but employees are allowed to be stockholders and have more freedom to choose the mix of com-ponents that make up their package Team innovation awards Competency-based awards
Other HRM Practices	Low employee participation Implicit job analysis Job enrichment	High employee participation Implicit job analysis Job enrichment Cross-functional teams communication: feedback on new product sales

Source: Daniel Jime' nez-nez and Raquel Sanz-Valle.

about the importance of linking HRM and innovation and, second, regarding the form of some HRM practices, particularly the use of external sources of recruitment, performance appraisals and incentives. There is no consensus regarding other HRM practices such as employment security, training, career paths or employee participation. Furthermore, empirical research conclusions are also heterogeneous. For this reason, can be formulated propositions about the relationship between innovation and HRM practices following both Miles and Snow (1984) and Schuler and Jackson (1987).

The first hypothesis refers to isolated HRM practices.

1. The strategy developed by a firm determines the HRM practices it carries out.

Thus, innovative firms will carry out HRM practices consistent with this strategy.

This hypothesis is broken down into two according to the theoretical model employed:

a. Firms following an innovation strategy will be characterized by the use of external sources of recruitment, low employment security, narrow application of training, very little use of internal career paths, the use of performance appraisal systems, incentive-based compensation and low employee participation.

b. Firms following an innovation strategy will be characterised by the use of external sources of recruitment, high employment security, broad application of training, the use of internal career paths, the use of performance appraisal systems, incentive-based compensation and high employee participation.

The second hypothesis is formulated from a configurational perspective:

2. The strategy developed by a firm determines the HRM system it implements.

Thus, innovative firms will implement an HRM system consistent with this strategy.

As in the previous case, alternative hypotheses are formulated according to the reference models, Miles and Snow (1984) or Schuler and Jackson (1987):

a. Firms following an innovation strategy will adopt an HRM system characterised by the use of external sources of recruitment, low employment security, narrow application of training, very little use of internal career paths, the use of performance appraisal systems, incentive-based compensation and low employee participation.

b. Firms following an innovation strategy will adopt an HRM system characterised by the use of external sources of recruitment, high employment security, broad application of training, the use of internal career paths, the use of performance appraisal systems, incentive-based compensation and high employee participation.

The assumption that strategy determines the firm's HRM practices is implicit in these hypotheses, as most of the contingent literature suggests.

3. The HRM practices developed by a firm determine its strategy.

Thus, firms that carry out HRM practices consistent with innovation will follow an innovation strategy.

a. Firms characterised by the use of external sources of recruitment, low employment security, narrow application of training, very little use of internal career paths, the use of performance appraisal systems, incentive-based compensation and low employee participation will seek a more innovative strategy.
b. Firms characterised by the use of external sources of recruitment, high employment security, broad application of training, the use of internal career paths, the use of performance appraisal systems, incentive-based compensation and high employee participation will seek a more innovative strategy.

Regarding the configurational approach, the following hypotheses are formulated:

4. The HRM system developed by a firm determine its strategy.

Thus, firms that implement an HRM system consistent with innovation will follow an innovation strategy.

a. Firms adopting an HRM system characterised by the use of external sources of recruitment, low employment security, narrow application of training, very little use of internal career paths, the use of performance appraisal systems, incentive-based compensation and low employee participation will seek a more innovative strategy.
b. Firms adopting an HRM system characterised by the use of external sources of recruitment, high employment security, broad application of training, the use of internal career paths, the use of performance appraisal systems, incentive-based compensation and high employee participation will seek a more innovative strategy.

Some European studies based on the models mentioned above, on small and medium enterprises, revealed the relation between HRM and innovation. According to the literature, HRM is a key element for the success of innovation, not many empirical studies have provided support for it. Using a contingency approach, different strategies will require different employee skills, knowledge and behaviours to be implemented. Presumably, HRM policies can influence these employee characteristics. There is agreement between specialists about, first, the relationship between innovation and HRM and, second, that in order to improve innovation, the adoption of HRM bundles is superior to, any of the individual. HRM practices of which they are' composed. On the other hand, there are inconsistencies in the literature about the contents of these bundles. Some authors suggest that employment security, extensive training or employee participation have a positive impact on innovation, while others think they have the reverse effect.

10.5 Conclusions

Alternative hypotheses, following the two most widely-accepted theoretical models, have been proposed in this paper. The conclusions drawn from European studies about HRM impact on firms' innovations using the above models are the following;

- There are empirical evidences that innovation explains the adoption of some HRM practices. The choice of an innovation strategy implies the use of an incentive-based compensation, the encouragement of employee participation, the use of appraisal systems and the use of broad internal career opportunities.

- The HRM practices condition the firm's orientation towards Innovation. The participation and the use of promotion plans significantly explain the firm's innovation orientation.

The importance of aligning HRM practices and innovation is a clear implication for managers, Schuler and Jackson's model and Miles and Snow's explains the fit between innovation and HRM. The firms that adopt an innovation strategy are more likely to use internal labour markets than external ones. The use of HRM practices aimed at building a stable group of employees in the company, which can adopt risks and experiment and which can participate in the adoption of the decisions that affect their jobs. This is more likely to create the conditions for the emergence of the new ideas that feed innovation.

References

1 Carneiro Alberto, "What is required for growth?, "Business Strategy Series,Vol.8, No.1 2007, pp. 51–57.
2 Dalota Marias Dan, "Successful Implementation of Knowledge Management in Small and Medium Enterprises", Romanian Economic and Business Review Vol.6 No.1, 2011, pp. 7–18.
3 Danial Jime'nez-Jime'nez and Raquel Sanz-Valle, "Innovation and human resource management fit: an empirical study ", International Journal of Manpower Vol. 26 No. 4, 2005 pp. 364–381
4 Jackson, S.E. and Schiller, R.S., "Understanding human resource management in the context of organizations and their environments", Annual Review of Psychology, Vol. 46, pp.1995, 237–64.
5 Laursen, K., "The importance of sectorial differences in the application of complementary HRM practices for innovation performance ", International Journal of the Economics of Business, Vol. 9 No.1, 2002, pp. 139–56
6 Miles, R.E. and Snow, C.C., "Designing strategic human resources systems", Organizational Dynamics, summer, 1984, pp, 36–52
7 Porter, M.E., "The technological Innovation dimension of competitive strategy", in Rosenbloom, R.S. (Ed), Research in Technological Innovation, Management, and Policy, 1983, JAI Press, Greenwich, CT.

11

The Importance of Using Human Resources Information Systems (HRIS) and a Research on Determining The Success of HRIS

Yasemin Bal, Serdar Bozkurt and EsIn Ertemsir

Yildiz Technical University, Turkey

11.1 Abstract

With the increasing effect of globalization and technology, organizations have started to use information systems in various functions and departments in the last decades. Human resources management is one of the departments that mostly use management information systems. HR information systems support activities such as identifying potential employees, maintaining complete records on existing employees and creating programs to develop employees' talents' and skills. HR systems help senior management to identify the manpower requirements in order to meet the organization's long term business plans and strategic goals. Middle management uses human resources systems to monitor and analyze the recruitment, allocation and compensation of employees. Operational management uses HR systems to track the recruitment and placement of the employees. HRIS can also support various HR practices such as workforce planning, staffing, compensation programs, salary forecasts, pay budgets and labour/employee relations. In this research, HRIS perception and HRIS satisfaction questionnaires were applied to HR employees in order to assess the effectiveness and use of HRIS in organizations. 78 questionnaires were received from HR employees working in different sectors. The results of the research give valuable insights about the success and effectiveness of HRIS in organizations. Also the results of the study are discussed in the context of the theoretical and empirical background of MIS and HRIS.

Keywords: Management information systems, Human resources information systems.

11.2 Introduction

Organizations must treat information as any other resource or asset. It must be organized, managed and disseminated effectively for the information to exhibit quality. Within an organization, information flows in four basic directions as upward, downward, horizontal and outward/inward (Haag & Cummings, 2008). Taking into account that there is a huge amount of information flow in organizations, it will be possible to understand the importance of information systems in organizations.

The information systems field is arguably one of the fastest changing and dynamic of all the business processions because information technologies are among the most important tools for achieving business firms' key objectives. Until the mid-1950s, firms managed all their information flow with paper records. During the past 60 years, more and more business information and the flow of information among key business actors in the environment has been computerized. Businesses invest in information systems as a way to cope with and manage their internal production functions and to cope with the demands of key actors in their environments. Firms invest in information systems for the business objectives such as achieving operational excellence (productivity, efficiency, agility), developing new products and services, attaining customer intimacy and service, improving decision making, achieving competitive advantage and ensuring survival (Laudon & Laudon, 2009).

11.3 Management Information Systems (MIS)

It is important to coordinate and control major functions, departments and the business processes in an organization. Each of these functional departments has its own goals and processes and they obviously need to cooperate in order for the whole business to succeed. Firms achieve coordination by hiring managers whose responsibility is to ensure all the various parts of an organization work together. The hierarchy of management is composed of senior management which makes long term decisions, middle management which carries out programs and plans and operational management which is responsible for monitoring the daily activities of the business. Each of these groups has different needs for information given their different responsibilities (Laudon & Laudon, 2009). Management information system (MIS) is designed to assist managerial and professional workers by processing and disseminating vast amounts of information to managers' organization-wide (Alavi & Leidner, 1999).

Management information system supplies information for strategic, tactical and operational decision making to all subsystems within the organization. This information provides an essential part of the feedback control mechanism in these areas and is necessary for the realization of subsystem objectives (Curtis & Cobham, 2002). Management information system is any system that provides information for management activities carried out within an organization. The information is selected and presented in a form suitable for managerial decision making and for the planning and monitoring of the organization's activities.

11.4 Human Resources Information Systems (HRIS)

Along with the upsurge of computerized management information systems (MIS) in industrialized countries' enterprises in the 1980s, HR functions increasingly started to deploy human resource information systems in their daily work. HRIS were primarily seen as MIS sub functions within HR areas intended to support the "planning, administration, decision-making, and control activities of human resource management. During the 1990s, along with the adoption of more complex HR practices focused on a company's overall performance goal, HRIS correspondingly evolved into more sophisticated information expert systems featuring analytical tools to support decision-making in managing human capital (Ostermann, Staudinger & Staudinger, 2009). Information technology in the past decade drastically changed the human resources function. Providing support for mainly administrative activities such as payroll and attendance management in the beginning, information technology today enhances many of the recruitment function's sub processes such as long and short-term candidate attraction, the generation, pre-screening, and processing of applications or the contracting and on boarding of new hires. Online job advertisements on corporate web sites and internet job boards, online CV databases, different forms of electronic applications, applicant management systems, corporate skill databases, and IS supported workflows for the contracting phase are only few examples of the various ways by which information systems today support recruitment processes (Keim & Weitzel, 2009).

In HR planning process it is easier to follow workforce gaps, the quantity and quality of the labour force and to plan future workforce requirements with the help of HR knowledge systems (Dessler, 2005). HRIS can support long range planning with information for labour force planning and supply and demand forecast; staffing with information on equal employment, separations and applicant qualifications; and development with information on training programs, salary forecasts, pay budgets and labour/employee relations with information on contract negotiations and employee assistance needs (Shibly, 2011). Risk and security management is another crucial function which can be derived by HRIS by following private and highly sensitive individual data and multiplatform security aspects which are perhaps the most serious factors that need to be taken into consideration (Karakanian, 2000).

HRIS is defined as an "integrated system used to gather, store and analyze information regarding an organization's human resources' comprising of databases, computer applications, hardware and software necessary to collect, record, store, manage, deliver, present and manipulate data for human resources function" (Hendrickson, 2003). An HRIS can perform a number of functions from the simple storage and communication of information, to more complex transactions. As technology advances, the range of functions that an HRIS can undertake increases. Actually HRIS is directed towards the HR departmentit self (Ruel, Bondarouk & Looise, 2004), but the use of HRIS can provide a number of benefits not only to the HR function, but also line managers, and the wider organization (Parry, 2009). The use of HRIS has been advocated as an opportunity for human resource professionals to become strategic partners with top management. HRIS allow HR function to become more efficient and to provide better information for decision making (Beadles, Lowery & Johns, 2005).

The increased use of web technology to deliver HR will leave HR specialists more time for strategic decision making and that outsourcing of people-management activities will liberate HR specialists to perform more strategic activities (Kulik & Perry, 2008). According to Ulrich (2007; 2009) as one of the strategic partners, the HR manager derives benefit from IHRS, to disseminate and execute the strategy within the organization. These systems enable employees to manage much of their own HR administrative work. They can take care of many routine transactions whenever they wish, because, automated systems don't keep office hours. In addition to their former operational role, HR professionals can also act as a competency manager by arranging the right people to the right positions in the right time with their new strategic architecture role (Gurol, Wolff & Ertemsir, 2010). HRIS is thought to contribute to overall business performance by fulfilling or at least supporting the tasks of data storage and retrieval, of serving as primary administrative support tools, of reporting and statistics as well as of program monitoring (Ostermann, Staudinger & Staudinger, 2009). HRIS plays an important role for any organisation to effectively manage its human assets. Many organizations have adopted HRIS to assist their daily human resources operations. HRIS must align and satisfy the needs of the organization and its users in order to be successful (Noor & Razali, 2011).

However, HR departments need to recognize some of the current limitations of web technology and its integration to the HRIS backbone. Similar to most e-business ventures, security of private HR information is a top priority. Organizations looking seriously into internet enabling of their HR businesses should evaluate the authentication, security, access rules, and audit trails related to service providers' networks, servers, and applications (Karakanian, 2000). On the other hand there can be undesired and unexpected consequences of HRIS. Undesired consequences refer, for instance, to an increase of quantity but a decrease of quality of applicants in e-recruiting (Strohmeier, 2009). Another important aspect of using information systems is user satisfaction. It is often suggested as an indicator of IS success. Many IS empirical researchers have regarded user satisfaction as important proxy of IS success and it is the most employed measure of IS success due to its applicability and ease of use. Within this literature, system and information characteristics have been core elements on user satisfaction which is defined as the attitude that a user has toward an information system (Shibly, 2011).

11.5 Research Methodology

The purpose of the study is to determine the relationship between the satisfaction of employees from HRIS and their perceptions of HRIS. Another aim of the study is to reveal the perceptions of employees for the dimensions that constitute HRIS and explain the points that should be developed. Also, it is aimed to reveal that if the HRIS perceptions of employees show difference or not according to their demographic qualifications (age, gender, seniority, position, education). The research has importance to determine the contribution and success of using HRIS for HR employees.

With the results of the research, it is possible to give valuable insights about the importance of using HRIS and the satisfaction level of HR employees from this system. There is a lack of empirical study in the related literature. By considering the need of empirical studies in this field, it is obvious that both the theoretical and empirical results of this research will give an important contribution to the related literature.

11.5.1 Sample and Data Collection

HR employees from different sectors participated to the research (n = 78) between January–March 2012. The method of the research sampling is "purposive sampling" which gives the researchers to use their own judgment to select suitable people for the sample (Balci, 2004).

Two scales were used in the questionnaires as measurement instrument of the research. The first scale has 4 dimensions and 22 items that measure HRIS. The scale was translated to Turkish by the researchers and used as with 3 dimensions and 22 items in the research according to face validity. Face validity is a judgment by scientific community that the indicator really measures the construct (Neuman, 2004).

The second scale has 3 items that measures HRIS satisfaction. The questions about social demographic qualifications such as gender, education, seniority, position and age were included to the items for measuring HRIS and the questionnaire form was developed. The questionnaires were sent to employees via e-mail and collected by the same way. Questionnaires were sent to HR employees who are working in different sectors such as pharmacy, fast-moving consumer and banking. 78 HR employees replied the questionnaires from these sectors.

11.5.2 Data Analysis

The items of HRIS perception and HRIS satisfaction scales were presented using a five-point Likert item as "1: strongly disagree" and "5: strongly agree". Data was analyzed by SPSS for Windows 18.0 package program. Firstly, Kolmogorov-Smirnov test was used to determine the normality of data and the results showed that data was distributed normally and it is possible to make parametric tests.

Cronbach alpha reliability value was computed in order to find the reliability of the scale. The reliability values are 0.961 for HRIS perception scale and 0.829 for HRIS satisfaction scale. The reliability values of both scales are high for researches in social sciences (Kalayci, 2005). Descriptive statistical analysis (arithmetic mean and standard deviation) and Pearson correlation test were used to determine the relationship between HRIS perception and HRIS satisfaction. Also, t-test and one-way ANOVA test were used to determine the differences according to demographic qualifications.

11.5.3 Research Hypothesizes

In this research, the relationship between the satisfaction of employees from HRIS and their perceptions of HRIS are analyzed, furthermore it is tested whether or not these two variables differ according to demographic variables such as position, gender, education level and age.

The main hypothesizes of the research are given below;

H_1: There is a relationship between the satisfaction of employees from HRIS and their perceptions of HRIS.

H_{1a}: Employees' perceptions of HRIS show difference according to their position.

H_{1b}: Satisfaction of employees from HRIS shows difference according to their position.

H_{1c}: Employees perceptions of HRIS show difference according to their genders.

H_{1d}: Satisfaction of employees from HRIS shows difference according to their genders.

H_{1e}: Employees perceptions of HRIS show difference according to their education levels.

H_{1f}: Satisfaction of employees from HRIS shows difference according to their education levels.

H_{1g}: Employees perceptions of HRIS show difference according to their ages.

H_{1h}: Satisfaction of employees from HRIS shows difference according to their ages.

11.5.4 Findings and Results

Employees from different sectors participated to the research (n $=$ 78). The social demographic qualifications of the participants are as follows: 25 male (32,1 %) and 53 female (67,9 %) participated to the research. The educational background of the participants are; 9 people (11,5 %) have high school degree, 44 people (56,4 %) have undergraduate degree and 24 people (32,1 %) have graduate and Ph.D. degree.

Table 1   **Demographic Qualifications of Participants**

Variable Name		Frequency	(%)
Gender	Male	25	32,1
	Female	53	67,9
Education	High School	9	11,5
	Degree Undergraduate	44	56,4
	Graduate / Ph.D.	24	32,1
Age	18–25	8	10,3
	26–33	49	62,8
	34–41	13	16,7
	42 and more	8	10,3
Seniority	Less than 1	20	25,6
	1–3 year	18	23,1
	4–6 year	17	21,8
	7–9 year	9	11,5
	10 and more	14	17,9
Position	Manager/Vice Mngr.	34	43,6
	Specialist	36	46,2
	HR Assistant	6	7,7
	No answer	2	2,5

The age classification of participants are; 8 people (10,3 %) are between 18 25; 49 people (62,8 %) are between 26 33; 13 people (16,7 %) are-between 34 41 and 8 people (10,3 %) are more than the age of 42. The seniority of participants are; 20 people (25,6 %) have less than 1 year seniority; 18 people (23,1 %) have 1 3 year seniority; 17 people (21,8 %) have 4–6 year seniority; 9 people (11,5 %) have 7 9 year seniority and 14 people (17,9 %) have seniority more than 10 years.

The positions of participants are; 34 people (43,6 %) HR manager and HR vice manager; 36 people (46,2 %) HR specialist and 6 people (7,7 %) HR assistant, 2 people did not give answer about their positions.

According to the results of descriptive statistical analysis (arithmetic mean and standard deviation) belong to two scales; the arithmetic mean for HRIS perception scale was computed as 3,73 (SD: 0.67) and the arithmetic mean for HRIS satisfaction scale was computed as 3,66 (SD: 0.78). These scores show that the participants gave answers to both scales as "agree". According to the arithmetic mean of the dimensions of HRIS perception scale, we see that the three dimensions of the scale are evaluated close to each other and participants perceive HRIS activities almost with the same importance (Table 2).

When we look at the arithmetic mean of the dimensions of HRIS satisfaction scale, we see that it has been perceived almost same as HRIS (arithmetic

Table 2 Descriptive Statistics Results of the Scales and Their Dimensions

Scales and Dimensions	Mean	Std. Dev.	N
System Quality	3,68	,69	78
Information Quality	3,76	,70	78
Perceived Ease of Use	3,73	,78	78
HRIS (total)	3,73	,67	78
HRIS Satisfaction (total)	3,66	,78	78

Table 3 Correlation Values of Relationship Between HRIS Dimensions and Satisfaction from HRIS

	System Quality	Information Quality	Perceived Ease of Use	HRIS Satisfaction
System quality	1			
Information Quality	,798**	1		
	,000			
Perceived Ease of	,770**	,766"	1	
Use	,000	, 000		
HRIS Satisfaction	,774**	,792**	,765**	1
p	,000	,000	,000	

**p<0.01

mean: 3,66). Here we see that participants emphasize the importance of HRIS and at the same time they are satisfied with the HRIS system they use.

H_1: There is a relationship between the satisfaction of employees from HRIS and their perceptions of HRIS.

Pearson correlation analysis was used to test the relationship the satisfaction of employees from HRIS and their perceptions of HRIS. According to the results of Pearson correlation analysis, a positive and high level relationship was found with Pearson correlation value: 0,841 ($p < 0.01$). H_1 hypothesis is accepted.

In order to understand this positive relationship better, also the relationship between HRIS dimensions and satisfaction from HRIS were investigated as shown in Table 3. According to correlation analysis results, positive and high level relationship was found among all dimensions of HRIS and system quality, information quality and perceived ease of use.

H_{1a}: Employees' perceptions of HRIS show difference according to their position.

In order to determine whether employees' perceptions of HRIS show difference or not according to position variable, t test was used and statistically meaningful difference was found between these two variables (p; 0.008 < 0,05). H_{1a} hypothesis is accepted.

Table 4   Results of One-way ANOVA Analysis for HRIS

		Sum of Squares	df	Mean Square	F	Sig.
HRIS	Between Groups	4,376	2	2,188	5,189	,008
	Within Group	30,781	73	,422		
	Total	35,157	75			

Table 5   Results of One-way ANOVA Analysis for HRIS Satisfaction

		Sum of Square	df	Mean Square	F	Sig.
HRIS	Between Groups	4,091	2	2,045	3,489	,036
Satisfaction	Within Groups	42,792	73	,586		
	Total	46,883	75			

Therefore we can say that the perception of employees working as a manager/ vice manager, specialist or HR assistant show difference. The arithmetic means of answers were investigated in the direction of difference. According to the results (manager/vice manager 3,92, specialist 3.67, assistant 3.03), the HRIS perception of managers and specialists are statistically and significantly different from the perceptions of HR assistants.

H_{1b}- *Satisfaction of employees from HRIS shows difference according to their position.*

In order to determine whether or not employees' satisfaction from HRIS show difference according to their position, one-way ANOVA test was used and statistically meaningful difference was found between these two variables (p: $0.036 < 0,05$). H_{1b} hypothesis is accepted.

Therefore we can say that the satisfaction of employees working as a manager/vice manager, specialist or HR assistant show difference. The arithmetic means of answers were investigated in the direction of difference. According to the results (manager/vice manager 3,85, specialist 3.58, assistant 3.00), the HRIS satisfaction of managers and specialists are statistically and significantly different from the satisfaction of HR assistants. (Table 5).

Also, t-test and one-way ANOVA test were used to determine whether or not perception of HRIS and satisfaction from HRIS show difference according to gender, education and age variables and no statistically meaningful difference was found between these variables. All analysis results of research hypothesizes are shown in Table 6.

Table 6   Results of Research Hypothesizes

Hypothesis	p-value	Result
H_1: There is a relationship between the satisfaction of employees from HRIS and their perceptions of HRIS.	**0.000**	**Accepted**
H_{1a}: Employees perceptions of HRIS show difference according to their position.	**0.008**	**Accepted**
H_{1b}: Satisfaction of employees from HRIS shows difference according to their position.	**0.036**	**Accepted**
H_{1c}: Employees' perceptions of HRIS show difference according to their genders.	0.238	Rejected
H_{1d}: Satisfaction of employees from HRIS shows difference according to their genders.	0.181	Rejected
H_{1e}: Employees' perceptions of HRIS show difference according to their education levels.	0.606	Rejected
H_{1f}: Satisfaction of employees from HRIS shows difference according to their education levels.	0.632	Rejected
H_{1g}: Employees' perceptions of HRIS show difference according to their ages.	0.146	Rejected
H_{1h}: Satisfaction of employees from HRIS shows difference according to their ages.	0.074	Rejected

11.6 Conclusion

HRIS is an integrated system used to gather, store and analyze information regarding an organization's human resources' comprising of databases, computer applications, hardware and software necessary to collect, record, store, manage, deliver, present and manipulate data for human resources function. The use of HRIS in organizations has various advantages for managers especially in decision making processes. In this study, the HRIS perception and HRIS satisfaction of HR employees were investigated. According to correlation analysis results, positive and high level relationships were found among all dimensions of HRIS and system quality, information quality and perceived ease of use which jointly constitute HRIS success. Also, it is found those employees' perceptions of HRIS show difference according to their position and satisfaction of employees from HRIS shows difference according to their position.

The results of the research reveal that HR employees perceive HRIS useful and they are satisfied with the system. It was found that both HRIS perception and HRIS satisfaction of employees show difference according to position variable. This finding may have its source from the limited access of HRIS functions depending on the positions of employees. Thus future studies should also consider the relationships between the access limitations to information content and functions of HRIS and user satisfaction. Overall present research provides valuable insights into the study of HRIS success.

References

1 Alavi, M., & Leidner, D. (1999). Knowledge management systems: Issues, challenges, and benefits, Communications of the Association for Information Systems, 1(7).

2 Balci, A. (2004). Sosyal bilimlerde arasturma: Yontem, teknik ve ilkeler, Ankara: Pegem.

3 Beadles, N. A., Lowery, C. M, & Johns, K. (2005). The impact of human resource information systems: An exploratory study in, the public sector, Communication of the IIMA, 5(4), pp. 39–46.

4 Curtis, G., & Cobham, D. (2002). Business information systems. London, UK: Pearson.

5 Dessler, G. (2005). Human Resource Management, 10. ed., USA: Prentice Hall

6 Gurol, Y., Wolff, A, & Ertemsir Berkin, E . (2010). E-HRM in Turkey: A case study, In I. Lee (Ed.), Encyclopedia of E-Business Development and Management in the Global Economy, pp. 530–540.

7 Hendrickson, A. R. (2003). Human resource information systems: Backbone technology of contemporary human resources, Journal of Labour Research, 24(3), pp.381–394.

8 Haag, S., & Cummings, M. (2008). Management information systems for the information age. New York, USA: McGraw Hill.

9 Karakanian, M. (2000). Are human resources departments ready for E-HR? Information Systems Management, 7 7(4), pp.1–5.

10 Kalayci, S. (Ed.). (2005). SPSS uygulamahli cok degiskenli istatistik teknikleri. Ankara: Asil Yayin Dagitim.

11 Keim, T., & Weitzel, T. (2009). An adoption and diffusion perspective on HRIS usage. In T. Coronas & M. Oliva (Ed), Encyclopedia of Human Resources Information Systems: Challenges in E~HRM(pp. 18–23). Hershey, PA: IG1 Global.

12 Kulik, C. T., & Perry, E. L. (2008). When less is more: The effect of devolution on HR's strategic Role and Donstrued Image, Human Resource Management, 47(3), pp.541–558.

13 Neuman, W. L. (2004). Basics of Social Research, USA: Pearson.

14 Noor, M. M., & Razali, R. (2011). Human resources information systems (HRIS) for military domain-a conceptual framework. International

Conference on Electrical Engineering and Informatics, 17–19 July, 2011, Indonesia.

15 Ostermann, H., Staudinger, B., & Staudinger, R. (2009). Benchmarking human resource information systems. In T. Coronas & M. Oliva (Ed,), Encyclopedia of Human Resources Information Systems: Challenges in E-HRM (pp. 92–101), Hershey, PA: IGI Global

16 Parry, E. (2009). The benefits of using technology in human resources management In T. Coronas & M. Oliva (Ed.), Encyclopedia of human resources information systems: Challenges in E-HRM (pp. 110–116). Hershey, PA: IGI Global.

17 Ruel, H., Bondarouk, T., & Looise, J. K. (2004). E-HRM: Innovation or Irritation. An Explorative Empirical Study in Five Large Companies on Web-based HRM, Management Revue, 75(3), pp. 364–381.

18 Shibly, H. (2011). Human resources information systems success assessment: An integrative model, Australian Journal of Basic and Applied Sciences, 5(5), pp. 157–169.

19 Strohmeier, S. (2009). Concepts of E-HRM Consequences: a Categorization, Review and Suggestion, International Journal of Human Resource Management, 20(3), pp.528–543.

20 Ulrich, D., Brockbank, W., Johnson, D., Sandholtz, K., & Younger, J. (2009). IK Yetkinlikleri. (Nazli Sahinbas Koksal, Trans,) Turkey/ Istanbul: Humanist Press. (Original Work- HR Competencies published 2008).

21 Ulrich, D. (2007). The New HR Organization. Workforce Management, 86(21), pp.40–44.

12

Human Resources' Development Needs in Higher Education

Osoian Codruta[1] and Zaharle Monica[2]

[1]*Senior Lecturer PhD, Babes-Bolyai University, Faculty of Economics and Business Administration, Management Department, Romania*
[2]*PhD Student Babes-Bolyai University Faculty of Sociology and Social Work, Sociology Department, Romania*

12.1 Abstract

An individual level approach of the human resource management requires the focus upon the employees' individual development within the organization. This process includes the identification of the human resources' development needs, the establishment of the methods and strategies necessary for developing the employees, the implementation of the training program, and the evaluation of the outcomes. The success of a development program strongly depends on the first stage, "which aims at analyzing the training needs. But, identifying employees' development needs within organization implies as well analyzing the organizational requirements and strategic objectives. The present study focuses on the faculty development needs identification stage, which includes: the identification of the professional objectives set by the academic staff, the establishment of the activities and strategies necessary for accomplishing the goals, and their comparison with the teaching dimensions assessed as deficient at the students' rating of instruction.

Keywords: Human resources development faculty development needs, Development management

Guided by the principles of a learning society, the European area increasingly values the development of its human resources. The awareness of the importance of human resources development upon the economic growth generated the design and implementation of several strategies at international level, in order to challenge the quality of the development programmes (Price, 2000).

At European level, there are a number of organizations with actions focused on the problematic of human resources improvement, which is thus rendered recognition as a catalytic element for the economic and social development. Since 1995, The CEI Working Group of Human Resources Development and Training has acknowledged the need of establishing a long-term collaborative programme with view to professional training (CEI, 2007). In this way, the Human Resources Development is recognized as a fundamental element of the economic and social development. Furthermore, the Action Plan for 2007–2009 is strongly focused on promoting lifelong learning and adult education. At the level of the OECD membership, the 21st century has revealed a series of new aspects connected with the Human Resources Management. The aging of masses phenomenon, along with further consequences of previous reforms, have triggered new initiatives of the OECD government with respect to the recruitment and maintenance of the highly qualified staff, to the human resources development programmes for meeting the needs within the work force, as well as to the development of the knowledge management (Shim, 2001). The growth of an organization is proportional to the development of the human resources working in it (Freeman, 1993). Tightly connected to the manner in which the management looks into the analysis of the immediate needs of the organization as well as into the long-term necessities in order to reach the established goals, the management of the human resources envisages a planning of the personnel, of its development and development needs. The performance-centered training focuses on both the individual and the organization (Brethower, Smelley, 1998): it develops the individual performance as well as it renders value to the whole organization. Therefore, the service of the training programmes is not anymore acknowledged only at micro level, but also at the macro level of the social economic organizations.

In the case of Romania, the meager finances for the human resources training programmes within public institutions led to an increase of the professional development needs. At present, there is an attempt to encourage such initiatives, by compelling employees in the medical and educational fields to accumulate credits by participating in these programmes. There frequently meet questions such as "Which of the employees should be trained?", "What type of training do they need?", and these can be answered by conducting development needs assessments. Consequently, the institutions that already offer or intend to implement training programmes need an assessment methodology for the human resources professional development needs.

The present study aims to put forward a diagnosis methodology for the training needs of teaching staff from higher education institutions, along as well as to present the conclusions drawn from the investigation of human resources development needs carried out within 4 universities. The literature suggests several models designed for identifying the development needs assessment. Starting from the traditional pattern proposed by McGehee and Thayer (1961), most of the available patterns in the field which have approached assessment of the training needs have taken into consideration 3 levels of analysis: the organizational, job analysis and personnel evaluation (Moore and Dutton, 1978). The unwinding of the training needs assessment process calls for gathering information (especially from reference persons within the institution), with respect to the optimum performance, real performance level, problem generating factors, solutions that reduce the discrepancy between the optimum and actual performance (Rossett, 1992).

As a rule, the study of the training needs reveal multiple opinions, which are usually generated by the diversity of organizational contexts (Szymanski, Linkowski, 1993). Differences with respect to the employees' training needs are identifiable according to the following variables: gender, schooling, hierarchy, work field, experience and type of organization. Moreover, the meta-analytic study conducted by Moseley and Heaney (1994) has revealed variables in what regards the assessment patterns and techniques depending on the working field. Besides factors such as the resources invested by organizations for enhancing the quality of their employees' activities, the value assigned by the institution to its human resources, or the colleagues' support (Foley, Redman, Horn, Davis, Neal, & Riper, 2003), the job performance also depends on the congruence between the content of the program and the specific needs of the employees. The professional development programs are more efficient when they meet the employees' knowledge and priorities (Cochran-Smith & Lytle, 1992). Understanding employees' development needs and their career objectives is a key element for an efficient development program (Evans & Chauvin, 1993). For this reason, an important step in designing a development program is to identify the faculty needs for professional improvement. In conclusion, the assessment and planning of the human resources development should also include the analysis of these characteristics.

The strategy suggested by Freeman (1993) inserts the following stages: 1) Meetings with different level organizational representatives; 2) Job analysis at this stage there is information gathering carried on about all the available positions in the institution; 3) Personnel assessment - the information obtained in the previous stage is processed in order to assess the employees' performance. The personnel evaluation offers information regarding the employees' weakness in their job performance; 4) Training and development assessment - this stage focuses on the assessment of the long-term training and development needs. The training needs depend on immediate needs and involve concrete changes in the immediate behavior of the staff, whereas the development needs hint at the acquisition of transversal abilities serviceable both in the present and in the future.

A distinction made in the scientific literature is that between needs assessment and needs analysis (Kaufman, Rojas, Mayer, 1993). If the needs assessment calls for the identification of the discrepancies between the actual and desired outcome and the establishment of the priorities in order to eliminate the shortcomings, the needs analysis focuses on the identification of the factors that have generated these discrepancies. Ignoring this distinction, most of the training needs assessment methodologies only track the needs as perceived by the employees.

Starting from the distinction between development needs and development wants, the purpose of this paper is to develop and propose a set of instruments which to facilitate the identification of the faculty' professional development needs in higher education. In order to overstep the bounds generated by the oblivion of the differences between the development needs and employees' wants, the methodology submitted within the present study is based on the pattern of discrepancies in the needs assessment (Kaufman, Rojas, Mayer, 1993) and purposes the analysis of the differences between the actual and wished for outcomes; this path calls for 2 stages.

The former, namely the development needs assessment, is a top-down strategy, oriented towards the future of the organization in which the primary objectives of the institution are being analyzed, in order to design the best training strategies which to concur to their fulfillment. Starting with the chief objectives of the institution, there will be a delimitation of the activities which will call for further information from the employees. Therefore, the employees' personal wishes can be diagnosed in an early stage of the process, so as the employees training needs compatible with the objectives of the institution are identified.

The second stage aims at the identification of the employees' perception upon the improvement of performances through training. Employees are asked to devise a top of the main factors that hinder them from optimum achievement of tasks, in a pre-investigation printed form. They assign training strategies to each identified problem. As opposed to the traditional methodology of needs identification, which has the respondents assign the type of training they consider necessary, this type of investigation compels the respondents to focus on the expected performances in their work.

The letter stage, namely the identification of opportunities for better performance, is a bottom-up strategy, which plans to involve all employees and diagnose the chief training needs. The focus is aimed at the development needs requisite for the performance improvement, as opposed to the training programmes desired by the employees. Thus, the conclusions submitted by the end of this assessment, approach the main training strategies in relation with the desired performances, in close connection with both the organization objectives and with the opportunities sighted by the employees in order to improve performance through training.

Four universities were part of the research sample: one large, public, general higher education institution (University 1); an average sized one (University 2); a rather large general higher education institution from the western part of the country (University 3); and a relatively recently established small private university (University 4). A total number of 570 questionnaires were filled in: 329 from the University 1; 148 from University 2; 42 from University 3; and 51 from University 4. The response rates ranged between 30 and 40%.

The assessment instrument of the development needs followed the identification of the relevant purposes for the faculty professional development and the identification of the activities believed to support the achievement of the main purposes.

The primary goal for the teaching staff from the four universities is: "the improvement of one's own line of specialization" (chosen as primary goal by 15% of the respondents), followed by "the improvement of teaching skills" (professed as main goal by 14% of respondents), and "participation in external scientific exchanges" (11% of the respondents). Comparing subgroups of faculty, the early career faculty members consider the development of their teaching skills slightly a more important goal. The participants holding the positions of junior lecturers tent to render higher importance to the improvement of teaching skills and intensification of interactivity with fellow faculty (ideas and experience exchange) than the professors.

The leading positions seem to be associated with differences among teaching staff in setting their professional goals. The teaching staff in leading positions renders the targets regarding the improvement of scientific research more importance than the other faculty (t = 2.09, p < .05). Depending on the general field/discipline of the teaching staff (sciences and social sciences and humanities), the independent samples /t test indicates differences in the goals established regarding the teaching skills, managerial, communication and exchange attendance. Thus, in opposition to the teaching staff from sciences, the social sciences faculty render significant more importance to the targets concerning teaching skills (t = 2.55, p < .05), managerial (t = 2.58, p < .05), communication skills (t = 2.24, p < .05), and mobility attendance (t = 2.02, p < .05).

The hierarchy of the strategies perceived as necessary for achieving the set goals is as follows: the access to prestigious scientific publications is chosen by 8.7% of the participants as the most appreciated activity, followed by the sessions of scientific events, such as congresses, symposiums, summer schools (8.3%), funds for participation in at least one annual conference (7.9%), work visits to foreign/local universities (7.8%), scientific partnership with other professors/researchers (7.7%), partnerships with the practitioners (7.2%).

The efficiency of the faculty development is greater if the content of the program addresses the specific needs expressed by the teaching staff. The present study identifies several differences in the development needs perceived by different subgroups of faculty, and for this reasons, development needs assessments might be extremely useful in designing and investing in successful faculty improvement programs, in order for the institutional development programs to distinctly address the faculty subgroups.

To sum up, the present paper proposes a set of instruments necessary for the identification of the professional development needs of the faculty in higher education. The analysis is based on the distinction between employees' development wants and needs, so that the methodology used to identity the necessary level as regards employees' development, according to the institutional strategic objectives and the specific characteristics of the teaching staff.

As regarding the limits of the present study, in order to immediately meet the organizational daily needs, one of the main difficulties is generated by the lack of organizational long term strategies and objectives. Another problem is represented by the lack of communication between the management and employees with regard the objectives establishment, which might lead to different set of goals for these two categories.

As the results emphasize, the efficiency of the development program is greater if the content of the program addresses the specific needs expressed by the employees. The present study identifies several differences in the development needs perceived by different subgroups of employees. Early career members are more oriented towards teaching skill improvement and development plans, while tenured faculty are rather interested in their research skills development. For this reasons, similar development needs assessments might be extremely useful in designing and investing in successful human resources improvement programs, in order for the institutional development programs to distinctly address the employees' subgroups.

References

1 Bates, Reid A., Holton, Elwood F., Naquin, Sharon S. (2000). Large-Scale Performance Driven Training Needs Assessment: A Case study, Public Personnel Management, Volume: 29, Issue: 2, pp. 253.
2 Brethower, D & Smalley, K. (1998). Performance-Based Instruction: Linking Training to Business Results. San Francisco: Jossey-Bass.
3 CEI (2007). Human Resource Development and Training, http://www.ceinet.org/main.php?pageID=61
4 Cochran-Smith, M., & Lytle, S. (1992). Inside-out: Teacher research and knowledge. New York: Teachers College Press,
5 Foley, B.J., Redman, R.W., Horn, E.V., Davis, G.T., Neal, E.M., & Van Riper, M.L. (2003). Determining Nursing Faculty Development Needs. Nurs Outlook, 51, 226–231.
6 Fowler, A. (1991), How to identify trening needs, Personnel Management Plus
7 Freeman, Jean M. (1993). Human Resources Planning - Training Needs Analysis, in Management Quarterly, Vol. 34
8 Kaufman, R, Rojas, A. M., & Mayer, H. (1993). Needs Assessment: A User's Guide. Englewood Cliffs, NJ: Educational Technologies Publications.
9 McGehee, W & Thayer, P. W. (1961). Training in Business and Industry. New York: Wiley.
10 McKillup, J. (1987). Need analysis; Tools for the human services and education. Newbury Park, CA: Sage.
11 Moore, M L. & Dutton, P. (1978). Training needs analysis. Academy of Management Review, 532–545.
12 Moseley, J L. & Heaney, M. J. (1994). Needs Assessment Across Disciplines. Performance Improvement Quarterly, 7, 60–79.
13 Nowack, K, M. (1991). A true training need analysis, Training and Development Journal, April

14 Pedhazur, E. J., & Schmelkin, L. P. (1991). Measurement, design, and analysis: An integrated approach. Hillsdale, NJ: Erlbaum.
15 Price, A. (2000). Principles of Human Resource Management: An Active Learning Approach, Blackwell Business
16 Rossett, A. (1992). Analysis of human performance problems. In H. Stolovich (Ed.) Handbook of Human Performance Technology (pp. 97–113). San Francisco: Jossey-Bass.
17 Shim, Deok-Seob (2001). Recent Human Resources Developments in OECD Member Countries, Personnel Management, Volume: 30, Issue:3, pp.32
18 Szymanski, Edna Mora, Linkowski, Donald C. (1993). Human resource development: An examination of perceived training needs of certified rehabilitation, Rehabilitation Counseling Bulletin, 00343552, Vol. 37, Issue 2.

13

Authors

1 Marta zuvic- Butorac, Faculty of Engineering
  Nena Roncevic, Faculty of Humanities and Social Sciences
  Damir Nemcanin, Faculty of Engineering
  Zoran Nebic, Faculty of Engineering
  University of Rijeka, Rijeka, Croatia
  martaz@riteh.hr; nrpncevic@ffri.hr;
  dnemcanin@riteh.hr; znebic@riteh.hr

2 Usman Sadiq,
  Ahmad Fareed Khan,
  Khurram Ikhlaq, Superior University, Pakistan
  Bahaudin G. Mujtaba, Nova Southeastern University, Pakistan
  Email: mujtaba@nova.edu

3 Samir R. Chatterjee
  President of the Society for Global Business
  and Economic Development.
  Professor of International Management
  Curtin University of Technology
  Australia
  Email: samir.chatterjee@cbs.curtin.edu.au

4 Ing. Iveta Gabcanova
  Univerzita Tomdse Bative Zline,
  Fakulta managements a ekonomiky,
  Mostni 5139, 760 01 Zlin,
  Ceska republika,
  Email: ivetagabcanova@gmail.com

5 Dr. Anita Trnavcevic
  University of Primorska, Slovenia

Anita.traavcevic@fm~kp.si
Dr. Nada Trunk Sirca
University of Primorska, Slovenia
Nada.trank@fm-kp.si
Mag. Vinko Logaj
National School for Leadership in Education, Slovenia
Vinko.logaj@solazaravnatelje.si

6 Franklyn Chukwunonso, francQnicostelo@vahoQ.com
234 8038765452, 8052829700
Department of Information Technology,
Federal University of Technology, Yola, Nigeria

7 Mary Ann Von Glinow,
John milliman,
Centre for effective Organization
Marshall School of Business
University of Southern California
Los Angeles, CA, USA

8 Roberto Biloslavo
Faculty of Management Koper, Slovenia
roberto.biloslavo@fm-kp.si
Max Zornada
Adelaide Graduate School of Business,
University of Adelaide,
Australia
mzornada@gsm.adelaide.edu.au
max.zornada@adelaide.edu.au

9 Angelos Alexopoulos
Kathy Monks
The learning Innovation and Knowledge Research
centre DCU Business School
Dublin City University
Glasnevin
Dublin 9,
Ireland

10 Professor Marius Dan Dalota, Ph.D.
Romanian-American University
1B, Expozitiei Avenue, Sector 1, Bucharest, Romania
dalota.marius.dan@profesor.rau

11 Yasemin Bal
Yildiz Technical University, Turkey
yaseminmutluay@gmail.com
Serdar Bozkurt
Yildiz Technical University, Turkey
serdarb21@gmail.com
EsIn Ertemsir
Yildiz Technical University, Turkey
esinertemsir@gmail.co

12 Osoian Codruta, Senior Lecturer PhD
Babes-Bolyai University
Faculty of Economics and Business Administration,
Management Department, Romania
codruta.osoian@econ.ubbcluj.ro
Zaharle Monica, PhD Student
Babes-Bolyai University
Faculty of Sociology and Social Work,
Sociology Department, Romania
monizaharie@staff.ubbcluj.ro

Appendices

Appendix A

Dear Participant,

We are the students of MSBA from Superior University, Lahore. We are conducting a research which aims to assess the Impact of Human Resource Information System (HRIS) on the HR Department in the banking sector of Pakistan.

This research is purely for academics purposes. It should not take more than 10 minutes to fill out this questionnaire. We are assuring you that all the information provided in this survey will be kept confidential and anonymous. Your cooperation in this regard is highly appreciated.

Name / Title:_____

_____(Optional)

Designation: _____
Occupation: 1. Private Job   2. Government   3.Business   4. Any other
Qualification: 1. M. Phil   2. Masters   3. Graduation

Scale: SA = Strongly Agree A Agree N=Neutral DA = Disagree SDA =Strongly Disagree

Sr No	Statements	SA5	A4	N3	DA2	SDA1
1	I am actively involved with the Human Resource Information System (HRIS)					
2	Overall I am satisfied with our HRIS.					
3	The employees of the Human Resources (HR) department appear to be satisfied with our HRIS.					
4	Overall we are satisfied with the modules we have installed and are available for use.					
5	Our HRIS has met our expectations.					
6	Our HRIS could be better utilized.					
7	Our HR employees understand how to use the Human Resource Information System.					
8	Our HRIS has made our HR decision-making more effective.					
9	Our HRIS has made the HR department more important to the institution.					
10	We are satisfied with the deployment of our HRIS.					
11	We are satisfied with the support we have received from our information technology (IT) department.					
12	We are satisfied with the support we have received from the software vendor.					
13	We are satisfied with the flexibility of the system.					
14	Our HRIS has decreased the time spent on recruiting.					
15	Our HRIS has decreased the time spent on training.					
16	Our HRIS has decreased the time spent on making staff decisions.					
17	Our HRIS has decreased the time spent on inputting data.					
18	Our HRIS has decreased the time spent on communicating information within our institution.					
19	Our HRIS has decreased the time spent on processing peaperwork.					
20	Our HRIS has decreased the time spent on correcting errors.					
21	Our HRIS has decreased cost per hire.					
22	Our HRIS has decreased training expenses.					
23	Our HRIS has decreased recruiting expenses.					

24 Our HRIS has decreased data input expense.
25 Our HRIS has decreased the overall HR Staff's salary expense.
26 Our HRIS has improved our ability to disseminate information.
27 Our HRIS has provided increased levels of useful information.
28 Our HRIS has improved increased levels of useful information.
29 Our HRIS has improved the training process.
30 Our HRIS has improved the data input process.
31 Our HRIS has improved the data maintenance process.
32 Our HRIS has helped with forecasting staffing needs.
33 Our HRIS decreased paperworks.
34 Our HRIS has increased security concerns.
35 The information generated from our HRIS is shared with top administrators.
36 The information generated from our HRIS is underutilized by top administrators.
37 The information generated from our HRIS has increased coordination between HR department and top administrators.
38 The information generated from our HRIS has added value to the institution.
39 The information generated from our HRIS helps our institution decide on employees raises.
40 The information generated from our HRIS helps our institution to make more effective promotion decisions.
41 The information generated from our HRIS helps our institution decide when to hire.
42 The information generated from our HRIS helps our institution make better decisions in choosing better people.
43 The information generated from our HRIS helps our institution decide when training and skill development are necessary.
44 Overall our administration thinks that HRIS is effective in meeting strategic goals.
45 The information generated from our HRIS has improved the strategic decision making of top administrators.
46 The information generated from our HRIS has made HR a more strategic partner in the institutioin.
47 Our HRIS has promoted our institution's competitive advantage.

Appendix B

HR Practices (18 items)

Selection and socialization (3 items, a= 70)

New employees are typically hired, based on their fit with the company's culture.

My company selects highly skilled and competent individuals to new posts.

As a new employee, I was encouraged to take part in company-sponsored social activities.

Quantity of training and development (2 items, a=84)

My company provides me with a well organised training and development programme.

My company allocates a generous amount of time and resources for my training and development needs.

Type of training and development (4 items, a= 68)

My training involves cross-functional group training and team building.

My training involves developing work-related social relationships with other employees across different areas, of my company.

Mentoring is an important development tool in my company.

Much of my training is on the job.

Performance Appraisal (2 items)[4]

My work performnce is evaluated based on the results of my team or work unit.

My work perforrnance targets are jointly determined by my manager and my team or work Unit members.

Rewards Mix (3 items, a=82)

Rewards are closely linked to my individual performance.

Rewards are closely linked to my team's/group's performance.

My company rewards me for sharing information and/or advice with my colleagues.

Rewards *competitiveness (2 items,* a = 89)

The pay levels in my company are relatively high compared to other firms in the industry.

The pay levels in my work unit are relatively high compared to other films in the industry.

Rewards equity (2 items, a=81)

There are small pay differences among the people in my work unit.

There are small pay differences across the various work units in my company.

Index